MATH ADVENTURES

A Key to Academic Math Advancement

GRADE 4

Author: Ace Academic Publishing

Ace Academic Publishing is a leading supplemental educational workbook publisher for grades K-12. At Ace Academic Publishing, we realize the importance of imparting analytical and critical thinking skills during the early ages of childhood and hence our books include materials that require multiple levels of analysis and encourage the students to think outside the box.

The materials for our books are written by award winning teachers with several years of teaching experience. All our books are aligned with the state standards and are widely used by many schools throughout the country.

Prepaze is a sister company of Ace Academic Publishing. Intrigued by the unending possibilities of the internet and its role in education, Prepaze was created to spread the knowledge and learning across all corners of the world through an online platform. We equip ourselves with state-of-the-art technologies so that knowledge reaches the students through the quickest and the most effective channels.

For inquiries and bulk orders, contact Ace Academic Publishing at the following address:

Ace Academic Publishing
3031 Village Market Place,
Morrisville, NC 27560, USA

www.aceacademicpublishing.com

This book contains copyright protected material. The purchase of this material entitles the buyer to use this material for personal and classroom use only. Reproducing the content for commercial use is strictly prohibited. Contact us to learn about options to use it for an entire school district or other commercial use.

ISBN: 978-1-949383-59-1
© Ace Academic Publishing, 2023

Introduction

Welcome to "**Math Adventures - A Key to Academic Math Advancement**"! This workbook is specifically designed to align with the school curriculum and help students improve their analytical and logical thinking skills. With over **750 questions and several word problems**, this book aims to cover all the required syllabus for students in Grade 4.

Our workbook is an excellent resource for end-of-the-year state tests given by schools, as well as a great review book during the summer. Whether you are looking to improve your math skills or simply keep them sharp, "**Math Adventures**" provides a comprehensive and challenging set of problems to help you achieve your goals.

Our authors have extensive experience in teaching and developing math curricula for students at all levels. **They have carefully crafted each problem to challenge students and help them develop key problem-solving and critical thinking skills.** The book covers a wide range of topics, including arithmetic, algebra, geometry, and data analysis, providing students with a well-rounded education in math.

We believe that with practice, anyone can master math. "**Math Adventures**" is designed to help students build confidence in their abilities and develop a love for the subject. With clear explanations, helpful hints, and detailed solutions, this book is an excellent tool for anyone looking to improve their math skills.

Thank you for choosing "**Math Adventures - A Key to Academic Math Advancement**".
We hope that you find it useful and enjoyable!

Common Core Math Workbooks

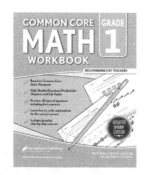 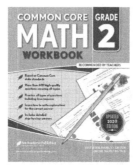  

  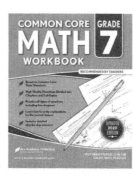 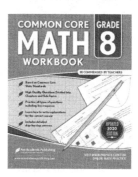

Common Core English Workbooks

The One Big Book Workbooks

   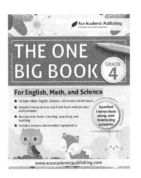

Math Adventures Workbooks

Early Learning Workbooks

TABLE OF CONTENTS

CHAPTER 1. MULTIPLICATION AND DIVISION — 1
1.1 Multiplication Properties and Facts — 2
1.2 Understanding Division — 9
1.3 Comparison of Multiplication and Division — 15
1.4 Equations for Multiplication and Division — 21
1.5 Chapter Review — 27

CHAPTER 2. PLACE VALUE — 33
2.1 Expanded Form and Numerals — 34
2.2 Friendly and Benchmark Numbers — 41
2.3 Multiply and Divide by Multiples of 10 — 46
2.4 Chapter Review — 51

CHAPTER 3. FACTORS AND PATTERNS — 57
3.1 Relationship Between Factors and Multiples — 58
3.2 Predict and Extend Growing and Repeating Patterns — 63
3.3 Pattern Rules — 69
3.4 Prime and Composite Numbers — 75
3.5 Chapter Review — 80

TABLE OF CONTENTS

CHAPTER 4. FRACTIONS   **85**

4.1 Equivalent Fraction Using an Area Model   86

4.2 Equivalent Fraction Using a Length Fraction Model   93

4.3 Add and Subtract Fractions   98

4.4 Mixed Numbers   104

4.5 Chapter Review   109

CHAPTER 5. CONVERSIONS   **117**

5.1 Time Across the Hours   118

5.2 Relating Conversions to Place Value   122

5.3 Centimetres and Meters   127

5.4 Grams and Kilograms   132

5.5 Litres and Millilitres   137

5.6 Chapter Review   143

CHAPTER 6. GEOMETRIC MEASUREMENT   **147**

6.1 Area of Rectilinear   148

6.2 Fixed Area - Varying Perimeter   155

6.3 Fixed Perimeter - Varying Area   160

6.4 Perimeter and Area of Rectangles   166

6.5 Chapter Review   171

TABLE OF CONTENTS

CHAPTER 7. REPRESENT AND INTERPRET DATA — **179**
7.1 Bar Graphs and Frequency Tables — 180
7.2 Line Plots and Categorical vs. Numerical Data — 193
7.3 Chapter Review — 203

CHAPTER 8. MEASURING ANGLES — **213**
8.1 Types of Angles and Lines — 214
8.2 Measuring Angles Using a Protractor — 219
8.3 Finding Unknown Angles — 226
8.4 Chapter Review — 232

CHAPTER 9. GEOMETRY — **239**
9.1 Angles and Sides of Quadrilaterals and Triangles — 240
9.2 Parallel and Perpendicular Lines of Quadrilaterals and Triangles — 247
9.3 Lines of Symmetry — 252
9.4 Chapter Review — 257

COMPREHENSIVE ASSESSMENT – 1 — **261**
COMPREHENSIVE ASSESSMENT – 2 — **275**
ANSWER AND EXPLANATION — **289**

CHAPTER 1
MULTIPLICATION AND DIVISION

MULTIPLICATION AND DIVISION

1.1 Multiplication Properties and Facts

Multiplication Properties and Facts

Multiplication Properties

What happens when you multiply a number by zero?

We must know that the product of any number and zero is zero. This is called the Multiplication Property of Zero.

Example: 0×5=0, 0×9=0, 0×0=0 etc.

What happens when you multiply a number by one?

We must know that multiplying any number by one equals the same number. We call this fact the Identity Property of Multiplication, and 1 is called the multiplicative identity.

Example: 1·5=5, 1×9=9, 1×1=1 etc.

We must know that the order in which two numbers are multiplied is not important.

This property of multiplication is called the commutative property.

Example: 2×3=6 and 3×2=6, 8×2=16 and 2×8=16 etc.

We must know that, when a product involves three or more factors, it does not matter which two factors are multiplied first.

This property of multiplication is called the associative property.

Example: 2×3×4=24 and 4×2×3=24

Thus, both can be written as the distributive property of multiplication:

This property states that, when a factor is multiplied by the sum of two or more numbers, we can multiply the factor by each of the numbers and then add.

Example: 5·2+3=5×2+5×3=10+15=25

MULTIPLICATION AND DIVISION

Multiplication Properties and Facts — 1.1

Multiplication Properties and Facts

Of course, adding first and then multiplying would yield the same result: 55 is equal to 25.

MULTIPLICATION FACTS

Multiplication is adding a number to itself a number of times. The symbol is used to denote multiplication.

For example, 6+6+6+6+6=30 is expressed using the multiplication symbol as
$$5 \times 6 = 30$$

5×6=30 is read as "five times six." The answer is 30. It is called the product.

Consider this multiplication example: 5×6=30

Each of the numbers 5 and 6 is called a factor, and the result of the multiplication is called the product.

$$\underset{\text{factor}}{5} \times \underset{\text{factor}}{6} = \underset{\text{factor}}{30}$$

Some of the words that indicate multiplication are given in the following table:

Operation	Word Phrase	Example	Expression
Multiplication	Times	3 times 7	3 x 7
	Product	The product of 5 and 3	5 x 3
	Twice	Twice 3	2 x 3
	Doubled	Doubled 6	2 x 6
	Tripled	Tripled 4	3 x 4

MULTIPLICATION AND DIVISION

1.1 Multiplication Properties and Facts

1 Sophia uses 21 beads for one bracelet she creates. How many beads does she need for 15 bracelets?

(A) 45 (B) 1956 (C) 243 (D) 315

2 There are 2,450 students at Rosewood Middle School. Half of the students buy lunch. Lunch costs $3. How much do the students at the school spend on lunch?

_____

_____

_____

3 A basketball team plays 18 games in a season. In each game, they score between 23 and 38 points. Which number could represent the total points the team scores in a season?

(A) 684 (B) 612 (C) 545 (D) 589

MULTIPLICATION AND DIVISION

Multiplication Properties and Facts — 1.1

4. There are 120 people going to a movie theater. One-third of the people are adults. The cost of an adult's ticket is $20, and the cost of a child's ticket is $15. How much will this group of people spend at the movie theater?

_____

_____

_____

5. There are 32 students in the 5 third grade classes. Which equation could be used to find the total number of students in the third grade?

- A) $(30 \times 11) - (2 \times 5)$
- B) $(28 \times 10) + (1 \times 15)$
- C) $(32 \times 5) + (2 \times 5)$
- D) $(28 \times 20) - (9 \times 15)$

6. Skip counting by five 7 times produces the same answer as skip counting by seven 5 times. This is true because

- A) $5 \times 7 = 35$
- B) $5 \times 7 = 7 \times 5$
- C) $5 + 7 = 7 + 5$
- D) $5 \times 7 = 5 \times (4+3)$

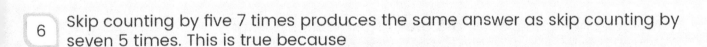

MULTIPLICATION AND DIVISION

1.1 Multiplication Properties and Facts

7. Ethan made apple juice on a hot summer day. He made 5 trays of apple juice. Each tray makes 12 glasses of apple juice. Use the distributive property to find the total number of apple juice.

A) 60   B) 130   C) 90   D) 140

8. Olivia makes $4 an hour. How much money does she make in a whole week, if she works 5 hours every day of the week?

A) $79   B) $140   C) $110   D) $21

9. A black chair costs $7 and a grey chair costs $7. Noah bought 5 black and 4 grey chairs. How much money does he spend?

A) $82   B) $40   C) $63   D) $25

MULTIPLICATION AND DIVISION

Multiplication Properties and Facts 1.1

10. There are 4 crates of mangoes, each containing 13 mangoes, and 8 crates of apples, each containing 13 apples. How much fruit is there in all?

    A) 126 B) 99
    C) 133 D) 156

11. Ava bought 9 boxes of pens with 12 pens in each box. Find the total number of pens.

    A) 112 B) 108 C) 102 D) 121

12. A rectangle has 4 sides. Altogether, how many sides do 11 rectangles have?

    A) 32 B) 24 C) 54 D) 44

13. Using multiplication, find the total number of dogs in 5 cages if each cage contains 3 dogs.

    A) 12 B) 9 C) 15 D) 18

MULTIPLICATION AND DIVISION

1.1 Multiplication Properties and Facts

14. Write a multiplication sentence for the total number of oranges in 9 boxes if each box contains 18 oranges. Find the missing number by solving.

   A) 9×18=162
   B) 9×18=152
   C) 9×18=172
   D) 9×18=192

15. A shopkeeper packs 8 sandwiches in a box. How many sandwiches does he pack in 12 boxes?

   A) 72
   B) 96
   C) 102
   D) 110

NEXT CHAPTER:
1.2 Understanding Division

MULTIPLICATION AND DIVISION

1.2 Understanding Division

Understanding Division

Division is sharing an amount equally.

Example: Sandy wants to share her bag of sweets with three friends. There are 6 sweets in the bag. How many sweets will each of them receive?

If 6 pieces of candy are shared equally among 3 children, each would get 2 pieces.
We write this as a division sentence: 6 ÷ 3 = 2.
This reads: "Six divided by three is two."
The symbol ÷ is used to denote division.
In 6 ÷ 3 = 2, 6 is called the **dividend**, 3 is called the divisor, and 2 is called the **quotient**.

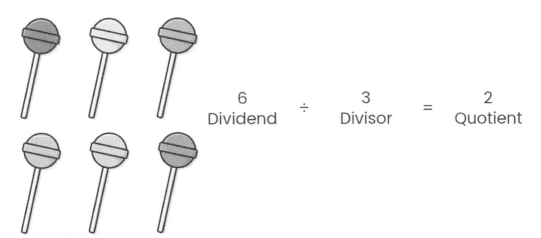

The operation of division helps us organize items into equal groups as we start with the number of items in the dividend and subtract the number in the divisor repeatedly.

MULTIPLICATION AND DIVISION

1.2 Understanding Division

1. A vegetable store has 752 tomatoes. There are 8 tomatoes in a package. Which strategy can be used to determine the total number of packages?

- A) $(744 \div 8) + 8$
- B) $(752 - 8) \div 8$
- C) $(700 \div 8) + (40 \div 1)$
- D) $(744 \div 8) + (8 \div 8)$

2. William picks 426 bananas and puts them all in baskets. There are 14 bananas in each basket. How many baskets does William use?

- A) 24
- B) 21
- C) 31
- D) 15

3. Four hundred and fifteen students are volunteering at the homeless shelter. There are 13 students on each team of volunteers. How many complete teams are there?

MULTIPLICATION AND DIVISION

Understanding Division 1.2

4 Mr. James collects $1,548 from ticket sales for the high school drama. Each ticket costs $9. How would you determine the number of tickets sold?

_____

_____

_____

5 Avery has collected 256 books, 86 notepads, and 189 pencils and puts them together in small bags. Each bag must contain 6 books, 2 notepads, and 4 pencils. How many complete bags will she create?

    Ⓐ 49   Ⓑ 42   Ⓒ 52   Ⓓ 47

6 Joseph's basketball team raises $14,520 selling jerseys and ball racks. Each jersey is sold for $8 and each ball rack is sold for $12. They sell 1,596 jerseys. How many ball racks does the team sell?

_____

_____

_____

MULTIPLICATION AND DIVISION

1.2 Understanding Division

7 To find the quotient of 16 and 4, one must look for

    (A) The product of 16 and 4.

    (B) A number is multiplied by 16, it gives 4.

    (C) When a number is divided by 16, the result is 4.

    (D) A number multiplied by 4 gives 16.

8 There are 104 people who want to eat lunch at a picnic. Each picnic table seats 13 people. Find the number of tables that the people need.

    (A) 8   (B) 6   (C) 7   (D) 9

9 Logan caught 81 lobsters at sea. Every hour, he caught nine lobsters. Find the number of hours Logan spent fishing.

    (A) 6   (B) 8   (C) 9   (D) 7

10 A dragon fruit costs $6. How many dragon fruits can be bought with $72?

    (A) 10   (B) 12   (C) 17   (D) 8

MULTIPLICATION AND DIVISION

Understanding Division 1.2

11. Amelia wants to equally share a basket of 105 carrots with 15 students. How many carrots will each student get?

   A) 7 carrots
   B) 9 carrots
   C) 12 carrots
   D) 17 carrots

12. The principal of the high school does not like the arrangement that Mr. John uses for bicycle storage. Describe another way the 64 bicycles can be arranged in rows with an equal number of bicycles in each row.

13. Mr. Isaac must order towels for all 27 players on his team.

   Each player gets 3 towels.
   Towels come in boxes of 9

   Find the number of boxes of towels he needs.

   A) 5
   B) 7
   C) 9
   D) 11

MULTIPLICATION AND DIVISION

1.2 Understanding Division

14 Ryan and Dylan each have the same number of pens. Ryan arranged his pens in 6 groups with 11 pens in each group. Dylan arranged his pens into 11 groups. How many pens did Dylan put in each group?

(A) 4 pens   (B) 5 pens   (C) 7 pens   (D) 6 pens

15 Hannah walks 4 km a day. After how many days would she have walked a total of 32 km?

(A) 6   (B) 8   (C) 9   (D) 7

NEXT CHAPTER:
1.3 Comparison of Multiplication and Division

MULTIPLICATION AND DIVISION

1.3 Comparison of Multiplication and Division

Comparison of Multiplication and Division

Multiplicative comparison means comparing two things or sets that need multiplication.

The division is an operation used to find the number of items in each group when the total number of items and the number of groups are known.

Example: A bottle can hold 3 cups of water. A bowl can hold 18 cups of water. How much more water can the bowl hold when compared to the bottle?

This is a division problem. The bowl can hold $18 \div 3 = 6$ times as much water as the bottle. We can translate comparative situations into equations with an unknown and then solve them.

Example: Ava has $45. She has five times as much money as Blake has. How much does Blake have?

Solution:

The amount of money Blake has is unknown. We can use a box to show Blake's money. Ava has 5 times more money than Blake.

We can see from the line drawing that we have to multiply Blake's money by 5 to get Ava's money.

Ava's money = Blake's money × 5. $45 = Blake's money × 5

Blake's money = $45 \div 5 = \$9$. Blake has $9.

MULTIPLICATION AND DIVISION

1.3 Comparison of Multiplication and Division

1. Kevin is 7 years old. Colton is 10 times as old as Kevin. Find Colton's age?

   A) 60   B) 65   C) 70   D) 75

2. Robert paid $15 for notebooks and he paid 5 times as much for school fees. How much did Robert pay for school fees?

   A) $ 70   B) $ 75   C) $ 78   D) $ 81

3. Parker paid $26 for a uniform. This was 2 times as expensive as a pair of shoes. What is the cost of the pair of shoes?

   A) $ 9   B) $ 11   C) $ 13   D) $ 15

4. Eva and her brother Adam are saving money to go to a national park. This month, Eva saved four times as much as she saved last month and Adam saved three times as much as he saved last month. Last month, Eva saved $11 and Adam saved $7. How much money did Adam save this month?

   A) $ 19   B) $ 21   C) $ 25   D) $ 27

MULTIPLICATION AND DIVISION

Comparison of Multiplication and Division 1.3

5. There are 66 dogs and 11 cats in the pet shop. How many times more dogs are there than cats?

   (A) 6   (B) 11   (C) 9   (D) 13

6. A monkey weighs 72 kg. A rabbit weighs 9 kg. How many times does a monkey weigh the same as a rabbit?

   (A) 2   (B) 4   (C) 8   (D) 16

7. A television costs $250. A smartphone is three times the price. How much does a smartphone cost?

   (A) $ 550   (B) $ 650   (C) $ 700   (D) $ 750

MULTIPLICATION AND DIVISION

1.3 Comparison of Multiplication and Division

8 Molly was doing a report on different animal weights. She made the chart below to show the weights of certain animals.

Animal	Dog	Sheep	Goat	Cow	Pig	Donkey	Rabbit
Weights	14 kg	25 kg	15 kg	140 kg	36 kg	80 kg	10 kg

What multiplication sentence relates to 10 and 140?

_____

_____

9 Choose from the following box a multiplication equation to match each comparison statement.

$45 = 5 \times 9$ (A) 45 days is 5 times longer than 9 days.

$52 = 13 \times 4$ (B) 9 pounds is 3 times as heavy as 3 pounds.

$50 = 5 \times 10$ (C) 52 inches is 13 times the length of 4 inches.

$9 = 3 \times 3$ (D) 50 balls is five times the number of ten balls.

MULTIPLICATION AND DIVISION

Comparison of Multiplication and Division   1.3

10. A cake shop offers cakes in five different sizes. The table below shows the price of a cake.

Cake Size	Price
Small	$4
Medium	$8
Large	$16
Extra-Large	$24
Jumbo	$32

How many times is a medium cake cheaper than a jumbo cake?

A) 4 times   B) 6 times
C) 7 times   D) 8 times

11. The fruit shop sells a variety of fruit. The table below shows the prices of some of them.

How many times is 1 kg of mango more expensive than 1 kg of banana?

A) 2 times   B) 4 times
C) 6 times   D) 1 time

Fruit	Price per kilogram
Apple	$24
Orange	$12
Mango	$36
Banana	$6

12. Jasmine's score on an English exam is 36 points. Naomi's score is 2 times as many points as Jasmine's score. How many points does Naomi have?

A) 32   B) 72   C) 82   D) 52

MULTIPLICATION AND DIVISION

Comparison of Multiplication and Division — 1.3

13. Mary has 40 red roses. Lilly has 6 times as many red roses as Mary. How many red roses does Lilly have?

    (A) 240 (B) 280 (C) 320 (D) 360

14. A white cloth is 90 meters long. It is six times longer than a pink cloth. How long is the pink cloth?

    (A) 11 (B) 13 (C) 17 (D) 15

15. Cora sold 28 pens. She sold seven times as many pens as Ivan. How many pens did Ivan sell?

    (A) 3 (B) 4 (C) 5 (D) 6

NEXT CHAPTER:
1.4 Equations for Multiplication and Division

MULTIPLICATION AND DIVISION

1.4 Equations for Multiplication and Division

Equations for Multiplication and Division

Equations can be built around three numbers, even if one of them is unknown. If we know that x stands for a number and x÷7=9, then other equations can be written relating the three numbers x, 7, and 9.

They are: x÷9=7, x=7×9 or x=97.

Example: Noah wants to cut a rope with a length of 18 inches into three equal parts. How long is each part? Complete the tape diagram and write an equation to solve.

Solution:

This is a division problem. The rope is to be divided into three equal parts.

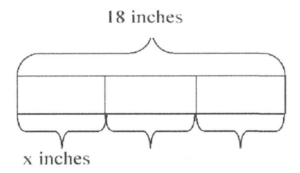

We can write an equation: 3×x=18

Therefore, x=18÷3, each part is 18÷3=6 inches long.

MULTIPLICATION AND DIVISION

1.4 Equations for Multiplication and Division

1 Given that x stands for a number and suppose that 63÷x=7 gives three other relations relating 7, x, and 63

- (A) x × 7 = 63
  7 × x = 63
  63 ÷ 7 = x

- (B) x − 7 = 63
  7 × x = 63
  63 ÷ 7 = x

- (C) 7 times

- (D) 8 times

2 The quotient of 54 and 9 is q. Write an expression that can be used to find the value of q. Use your expression to find the value of q.

- (A) 54+9=q
  q=63

- (B) 54÷9=q
  q=6

- (C) 54÷9=q
  q=12

- (D) 54−9=q
  q=12

3 Alexa travels 20 miles to get to the beach. Bella travels x miles to get to the beach. The equation 20÷x=4 can be solved to find the distance that Bella drives to the beach. Explain how the distance that Alexa drives to the beach and the distance that Bella drives to the beach compare. Find the distance that Bella drives to the beach.

- (A) 6 miles. Alexa drives two times as many miles as Bella

- (B) 3 miles. Alexa drives five times as many miles as Bella

- (C) 10 miles. Alexa drives seven times as many miles as Bella

- (D) 5 miles. Alexa drives four times as many miles as Bella

MULTIPLICATION AND DIVISION

Equations for Multiplication and Division — 1.4

4 Sarah sold 25 cups of green tea at her shop. Riley's shop sold 'n' cups of green tea. The equation 25−10=n gives the number of cups of green tea that Riley sold. Which of the following must be true?

- A) Riley sold 10 more cups than Sarah
- B) Riley sold 10 fewer cups than Sarah
- C) Riley sold 10 times as many cups as Sarah
- D) Sarah sold 10 times as many cups as Riley

5 A bakery has an order for 85 packs of bread. They decided to bake 5 packs of bread per hour. Write the equation to determine the number of hours (p) a bakery needs to finish the order.

- A) 85÷5=p
  p=17
- B) 85×5=p
  p=425
- C) 85−5=p
  p=80
- D) 85+5=p
  p=90

6 Lucy bought 4 boxes of cakes. There were 18 cakes in each box. How many cakes (y) did Lucy buy? Write an equation to solve.

- A) 4+18=y
  y=22
- B) 18−4=y
  y=14
- C) 4×18=y
  y=72
- D) 18÷4=y
  y=4.5

MULTIPLICATION AND DIVISION

1.4 Equations for Multiplication and Division

7 When a number is multiplied by 3 the result is 21. Use r for the number.

A) r×3=21
   r=6

B) r×3=21
   r=7

C) r×3=21
   r=8

D) r×3=21
   r=8

8 What multiplication expression can be used to find the number of triangles below?

A) 6×3=18

B) 3×4=12

C) 2×7=14

D) 5×3=15

9 The following expression can be used to describe the total number of emoji in the array below.

$$(2 \times z) \times 6$$

What is the value of z?

A) 1

B) 2

C) 3

D) 4

MULTIPLICATION AND DIVISION

Equations for Multiplication and Division — 1.4

10. Seven people earned money by selling baseball bats. They divided the money equally among themselves. Each person received y dollars. How much money did each of the seven people receive if their total earnings were 98?

- (A) y=54
- (B) y=34
- (C) y=14
- (D) y=44

11. Select the three equations that are true when the number 6 is put into each of the Question marks.

1. 5×?=30
2. 4÷?=4
3. 5×?=0
4. 48÷?=8
5. 2×?=12
6. 35÷?=5

- (A) 5×?=30
  48÷?=8
  2×?=1
- (B) 4÷?=4
  48÷?=8
  35÷?=5
- (C) 5×?=30
  48÷?=8
  2×?=12
- (D) 35÷?=5
  48÷?=8
  2×?=12

12. Rachel bough 11 boxes of pencils. Each box contains 12 pencils. Let r be the total number of pencils there are. Write an equation to determine r.
Find out how many pencils there are in total.

- (A) 112
- (B) 122
- (C) 132
- (D) 142

MULTIPLICATION AND DIVISION

1.4 Equations for Multiplication and Division

13 Victor is saving money to buy a new tablet that costs $600. He plans to save $60 each month. Determine the number of months Victor can use to, x, it will take him to save up for the tablet.

(A) 5   (B) 10   (C) 15   (D) 20

14 Alan bought and planted 9 trees in his garden. Each tree costs the same. He paid $45 for all nine trees. Let p represent the cost of each tree. Write an equation to determine p? Find the value of p.

(A) p=0   (B) p=1   (C) p=3   (D) p=5

15 Purchasing fifteen books costs $225. Use q for the cost of one book.

(A) q=35   (B) q= 25   (C) q=15   (D) q=5

NEXT CHAPTER:
1.5 Chapter Review

MULTIPLICATION AND DIVISION

1.5 Chapter Review

1. Mya ordered 12 packets of cookies. Each packet contains 36 cookies. How many cookies does Mya have?

2. A theater has 4 sections, and the same number of seats in each section. There were 1,752 people in the morning show. If the audience are assigned tickets evenly between the sections, how many people are in each section?

   A) 438   B) 452   C) 475   D) 493

3. There are 21 people in Mr. Bryan's family and 16 people in Mr. Steven's family. Each person in Mr. Bryan's family pays $25 for a trip to the aquarium. Each person in Mr. Steven's family pays $32 for a trip to the national park. Which family spends more money on their trip?

MULTIPLICATION AND DIVISION

1.5 Chapter Review

4 Leo has to type two pages. The first page has 24 lines, while the second has 18 lines. He takes 1 minute to type each line. What is the total time Leo spent typing two pages?

(A) 20   (B) 37   (C) 18   (D) 42

5 Mark sold 24 cupcakes in a day. Mateo sold 3 times as many cupcakes as Mark. How many cupcakes did Mateo sell on the same day?

(A) 24+3=27   (B) 24×3=72
(C) 24÷3=8   (D) 24−3=21

6 A school bus can take 40 students. If the total number of students is 8,000, how many school buses are needed?

(A) 100   (B) 150   (C) 200   (D) 250

MULTIPLICATION AND DIVISION

Chapter Review 1.5

7. A sugar cane garden contains 120,000 sugar cane plants. Each day, 500 sugar cane plants are cut to make sugar cane juice. Find the number of days it will take to cut all the sugar cane in the garden completely.

   (A) 240 days   (B) 260 days   (C) 340 days   (D) 360 days

8. 50 chocolates are distributed to ten children, ensuring that each child receives the same number of chocolates. Which operation gives the correct number of shares?

   (A) 50×10   (B) 50÷10   (C) 50+10   (D) 50−10

9. Emma bought 4 boxes of 20 water bottles each. Find the total number of water bottles in the boxes.

   (A) 80   (B) 60   (C) 40   (D) 20

10. Using multiplication, find the total number of books on 16 bookshelves if each shelf contains 30 books.

    (A) 180   (B) 280   (C) 380   (D) 480

MULTIPLICATION AND DIVISION

1.5 Chapter Review

11 There are 139 people working in the company. There are 15 employees on each team. How many complete teams are there?

12 Jacob drinks 3 liters of water per day. How many days total would it take to drink 69 liters of water?

Ⓐ 13   Ⓑ 19   Ⓒ 23   Ⓓ 27

13 There are 120 apple trees and 60 orange trees in the garden. How many times more apple trees are there than orange trees?

Ⓐ 1   Ⓑ 2   Ⓒ 3   Ⓓ 4

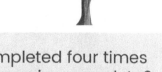

14 Rose solved seven exercises in the math book. Jasmine completed four times the number of exercises as Rose. How many exercises did Jasmine complete?

Ⓐ 28   Ⓑ 62   Ⓒ 89   Ⓓ 12

MULTIPLICATION AND DIVISION

Chapter Review 1.5

15 Given that n stands for a number and suppose that 121÷n=11 give three other relations relating 11, n, and 121.

- A) n+11=121
  11×n=121
  21÷n=11

- B) n−11=121
  11+n=121
  121−11=n

- C) n×11=121
  11×n=121
  121÷n=11

- D) n−11=121
  11−n=121
  121×11=n

16 A hexagon has 6 sides. Altogether how many sides do 14 hexagons have?

- A) 42
- B) 54
- C) 72
- D) 84

17 The building has 128 floors. There are 64 apartments on each floor. How many apartments are in the building?

_____

_____

18 Alice caught 36 fish. Mya caught x number of fish. The equation 36×3=x gives the number of fish Mya caught. Which of the following must be true?

- A) Mya caught three more fish than Alice.
- B) Mya caught three fewer fish than Alice.
- C) Mya caught three times as many fish as Alice.
- D) Alice caught three times as many fish as Mya.

31

MULTIPLICATION AND DIVISION

1.5 Chapter Review

19 Amy filled 7 boxes with pineapple. There were 12 pineapples in each box. How many pineapples, (p) did Amy use? Write an equation to solve.

- Ⓐ 12÷7=p
  p=1.7
- Ⓑ 12×7=p
  p=84
- Ⓒ 7+12=p
  p=19
- Ⓓ 12−7=p
  p=5

20 Mr. Antonio bought light bulbs for $579. Each light bulb costs $3. How would you determine the number of light bulbs to buy?

CHAPTER 2
PLACE VALUE

PLACE VALUE

2.1 Expanded Form and Numerals

Expanded Form and Numerals

In a large number, the first set of three digits, the units place, the tens place, and the hundreds place are together called the ones period. The second set of three places, the thousands place, the ten thousands place, and the hundred thousands place, are together called the thousands period. To make the number easier to read, we sometimes separate the digits in the thousands period from the digits in the ones period by a comma.

529,371

Thousands Period | Ones Period

529,371					
Thousands Period			Ones Period		
Hundred Thousands	Ten Thousands	Thousands	Hundreds	Tens	Units
5	2	9	3	7	1

Example: Read the number 197,048. Give the place value of each of the digits of the number.

Solution: One hundred ninety-seven thousand, forty-eight.

PLACE VALUE

Expanded Form and Numerals — 2.1

Expanded Form and Numerals

197,048					
Thousands Period		Ones Period			
Hundred Thousands	Ten Thousands	Thousands	Hundreds	Tens	Units
1	9	7	0	4	8

We have shown that one hundred ninety-seven thousand, forty-eight can be written as 100,000 + 90,000 + 7,000 + 40 + 8.

One hundred ninety-seven thousand, forty-eight written in standard form is 197,048; the form 100,000 + 90,000 + 7,000 + 40 + 8 is called the **expanded form**.

Example:

Write the number seventy-nine thousand, four hundred and sixty-three in:

a. standard form
b. expanded form

Solution:

a. 79,463
b. 70,000 + 9,000 + 400 + 60 + 3

PLACE VALUE

2.1 Expanded Form and Numerals

1. A candy container contains 280,777 candies. If one thousand additional candies are added, which digit in 280,777 would change?

   A) 1　　　B) 0　　　C) 9　　　D) 3

2. What number has 5 tens of thousands, 1 fewer thousand than ten thousands, 3 more hundreds than thousands, 1 more tens than hundreds, and 1 more ones than ten thousands?

   A) 54,786　　　B) 54,999　　　C) 53,297　　　D) 52,654

3. What number has 8 tens of thousands, 1 more thousand than tens of thousands, 4 fewer hundreds than thousands, 3 fewer tens than hundreds, and 4 more ones than ten thousands?

   A) 89,987　　　B) 89,524　　　C) 88,123　　　D) 87,564

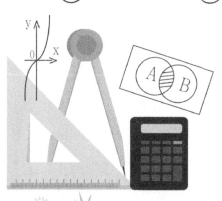

PLACE VALUE

Expanded Form and Numerals — 2.1

4 Vivian writes a number in her notebook with the following characteristics:

- 6 ten thousands
- 9 thousands
- 3 hundreds
- 1 tens
- 7 ones

What number does Vivian write in her notebook?

(A) 68,546   (B) 68,928   (C) 69,245   (D) 69,317

5 An aquarium contains 154,387 fish. If one thousand additional fish are added, which digit in 154,387 would change?

(A) 2   (B) 4   (C) 6   (D) 8

6 Write the number 2<u>4</u>,680 in expanded form and give the place value and value of the underlined digit.

(A) 20,000 + 4,000 + 600 + 80 + 0

The place value of the underlined digit in 2<u>4</u>,680 is 1,000s, while the value of that digit is 4,000.

(B) 20,000 + 5,000 + 600 + 80 + 10

The place value of the underlined digit in 2<u>4</u>,680 is 100s, while the value of that digit is 400.

PLACE VALUE

2.1 Expanded Form and Numerals

(C) 200,000+40,000+6,000+800+30

The place value of the underlined digit in 2<u>4</u>,680 is 1,000s, while the value of that digit is 6,000

(D) 20,000+4,000+600+100+10

The place value of the underlined digit in 2<u>4</u>,680 is 10s, while the value of that digit is 40.

7. Write a 6-digit number with the largest possible value, using each of the digits 1,7,2,5,8, and 6 exactly once. What is the place value of the digit 5 in the number you wrote?

(A) Ones (B) Tens (C) Hundreds (D) Thousands

8. Write a 5-digit number with the smallest possible value using each of the digits 9,5,3,6 and 1 exactly once. What is the place value of the digit 9 in the number you wrote?

(A) Thousands (B) Hundreds (C) Tens (D) Ones

9. Which number's underlined digit is worth 7,000?

(A) 15,1<u>7</u>9 (B) <u>7</u>7,987 (C) 4,6<u>7</u>7 (D) <u>7</u>6,976

PLACE VALUE

Expanded Form and Numerals — 2.1

10. Choose the option that shows the number 26,916 in words.

A) Twenty-six thousand nine hundred
B) Twenty-six thousand five hundred and ninety-six
C) Twenty-six thousand nine hundred and sixteen
D) Twenty-six thousand seven hundred and fifty-one

11. John saved a security code number with the following characteristics in his smartphone:

- 1 ten thousands
- 9 hundreds
- 5 tens
- 3 ones

What is the number that John has saved in his smartphone?

A) 10,953   B) 1,953   C) 19,053   D) 11,953

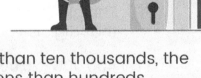

12. What number has 3 ten thousands, 2 fewer thousands than ten thousands, the same number of hundreds as ten thousands, 5 more tens than hundreds, and 4 more ones than ten thousands?

A) 37,254   B) 31,387   C) 34,367   D) 36,345

13. Which number's underlined digit is worth 70,000?

A) 65,31<u>7</u>   B) 12<u>7</u>,987   C) 24,6<u>7</u>0   D) 9<u>7</u>6,246

PLACE VALUE

2.1 Expanded Form and Numerals

14. James thought of a number that had the digit 9 in the ten thousands place and the digit 1 in the tens place. Which number was James thinking about?

- (A) 9,678
- (B) 192,713
- (C) 78,956vv
- (D) 999,563

15. Choose the number 421,98<u>3</u> in expanded form and give the place value and value of the underlined digit.

- (A) 2400,000+20,000+4,000+600+80+3

  The place value of the underlined digit in 421,98<u>3</u> is 1,000s, while the value of that digit is 1,000.

- (B) 40,000+1,000+600+80+10

  The place value of the underlined digit in 421,98<u>3</u> is 100s, while the value of the digit is 100.

- (C) 400,000+40,000+6,000+800+30

  The place value of the underlined digit in 421,98<u>3</u> is 1,000s, while the value of that digit is 1,000.

- (D) 400,000+20,000+1,000+900+80+3

  The place value of the underlined digit in 421,98<u>3</u> is ones, while the value of the digit is 3.

NEXT CHAPTER:
2.2 Friendly and Benchmark Numbers

PLACE VALUE

2.2 Friendly and Benchmark Numbers

Friendly and Benchmark Numbers

Benchmark numbers are numbers against which other numbers or quantities can be estimated and compared.
Benchmark numbers are usually multiples of 10 or 100.
Benchmark numbers are usually referred to as friendly numbers.
Benchmark fractions are fractions that are easy to picture mentally.
To get an estimate (a result close to the actual answer), a student rounds the numbers involved so that computation can be done mentally.
Estimate if you do not need the exact answer of a sum or a difference.
Round each number to the highest place value of the largest number, and then find the sum of the difference.

Example 1: Estimate, by rounding to the nearest 100, the sum: 642 + 823 + 811.
Solution: Rounding to the nearest 100

Round	Add	
600	600	The answer is about 2,200
800	800	
+800	+800	
	2,200	

Example 2: Estimate the difference 1,883 − 137 by rounding.
Solution:

	Round to the nearest 1,000	
1,883	2,000	The answer is about 2,000
− 137	− 0	
	2,000	

A better estimate can be obtained if we round to a lesser place value.

PLACE VALUE

2.2 Friendly and Benchmark Numbers

1. Estimate the sum 4,321 + 2,546

   (A) 7,000 (B) 6,000 (C) 8,000 (D) 5,000

2. Estimate 242 × 0.39. (Hint: 0.39 is close to $\frac{2}{5}$)

   (A) 20 (B) 40 (C) 80 (D) 100

3. Which two numbers give the same result when rounded to the nearest 100?

   (A) 587 and 478 (B) 527 and 478
   (C) 698 and 189 (D) 356 and 234

4. What is the largest whole number that gives 50 when rounded to the nearest 10?

   (A) 45 (B) 57 (C) 48 (D) 54

PLACE VALUE

Friendly and Benchmark Numbers — 2.2

5. What is the smallest whole number that gives 70 when rounded to the nearest 10?

- (A) 66
- (B) 69
- (C) 61
- (D) 74

6. Mark the following as True or False.
169 rounded to the nearest 100 is 200.

- (A) True
- (C) False

7. Daniel buys 31 watches. Each watch costs $16. Give an estimate for the total cost of the watches.

- (A) $610
- (B) $650
- (C) $600
- (D) $675

8. Which of the following is closest to 47×741?

- (A) 40,000
- (B) 35,000
- (C) 65,000
- (D) 70,000

43

PLACE VALUE

2.2 Friendly and Benchmark Numbers

9 Which of the following is closest to 526÷22

  A) 5   B) 15   C) 25   D) 35

10 Andrew saved $93,658 to buy a house. How do you round Andrew's savings to the nearest thousands?

  A) 94,000   B) 93,000   C) 95,000   D) 92,000

11 Grace bought some fruit at a shop. Her bill for the fruit she bought is shown in the table.

Fruit	Prize
Apple	$16
Orange	$11
Kiwi	$22
Pear	$19
Strawberry	$27

Estimate the total bill.

  A) $90   B) $150   C) $250   D) $100

PLACE VALUE

Friendly and Benchmark Numbers — 2.2

12. Sofia sold 398 balls. Each ball cost $7. Which of the following is closest to the total amount that Sofia received?

- A) $2,000
- B) $3,000
- C) $4,000
- D) $5,000

13. Mark the following as True or False.
425 rounded to the nearest is 500.

- A) True
- C) False

14. Estimate: 62,765÷541 (Hint round 62,765 to the nearest 1000)

- A) 95
- B) 101
- C) 112
- D) 126

15. Estimate the difference 325,654−23,851

- A) 280,000
- B) 270,000
- C) 260,000
- D) 250,000

NEXT CHAPTER:
2.3 Multiply and Divide by Multiples of 10

PLACE VALUE

2.3 Multiply and Divide by Multiples of 10

Multiply and Divide by Multiples of 10

The powers of 10 are: 10, 100, 1,000,...

We have seen from the multiplication table that when a number is multiplied by 10, the effect is to place a zero in the unit's place: 6×10=60 and 10×10=100.

A similar result can be concluded for multiplying a single digit by 100. Namely, place two zeros to the right of the digit: one for the tens place and one for the unit's place.

Example: 3×100=300; 15×100=1,500

When multiplying a single digit by 1,000, place three zeros to the right of the digit: one for the hundreds place, one for the tens place, and one for the unit's place.

Example: 8×1,000=8,000 and 4×1,000=4,000 etc.

Multiples of 10

A multiple of 10 is a whole number that ends with 0.

0, 10, 40, 80, 600, 800, 910 and 3,000 are all multiples of 10.

In fact, 600 and 800 are also multiples of 100 while 3,000 is a multiple of 1,000.

When multiplying by a multiple of 10, ignore the trailing zeros and do the multiplication. Place the zeros you ignored to the right of the last digit of your answer.

Example: Multiply 60×7.

Solution:

Ignore the zero in 60 and multiply 6×7.

6×7=42

Place the zero you ignored to the right of 2: the result becomes is 420.

Therefore, 60×7=420.

PLACE VALUE

Multiply and Divide by Multiples of 10 — 2.3

Multiply and Divide by Multiples of 10

Dividing Multiples of 10
Divide the front digits and write the quotient.
Cross out the zeros that match up from both numbers.
Example: $7{,}200 \div 90 = 720 \div 9 = (72 \div 9) \times 10 = 8 \times 10 = 80$.

1. Jackson works 4 hours every day. He works 25 days a month. What is the total number of hours he works in a month?

   (A) 200 hours (B) 150 hours (C) 100 hours (D) 50 hours

2. A box of balls contains 35 balls. Harris bought 7 boxes, and Ian bought 9 boxes. How many balls did both Harris and Ian buy?

   (A) 420 (B) 560 (C) 640 (D) 780

3. There are 250 rooms in a hostel. Each room has 6 students. How many students are there in 250 rooms?

   (A) 1,100 (B) 1,200 (C) 1,300 (D) 1,500

PLACE VALUE

2.3 Multiply and Divide by Multiples of 10

4 Max saves $50 every month for 10 months to buy a laptop. What is the cost of the laptop?

(A) $500 (B) $400 (C) $300 (D) $200

5 A bag contains 8 times as many chocolates as donuts. If there are 640 chocolates, what is the number of donuts in the bag?

(A) 70 (B) 80 (C) 90 (D) 100

6 X is a number, if you multiply X by 13 you get 7,540. What number is x?

(A) 925 (B) 430 (C) 675 (D) 580

7 Jack bought 500 flowers. The cost per rose is $6 and $10 for lily. What is the cost of the flowers he bought?

(A) $4,000 (B) $8,000 (C) $12,000 (D) $16,000

48

PLACE VALUE

Multiply and Divide by Multiples of 10 — 2.3

8 How many rings can Ivan buy for $4,200, if each ring costs $60?

(A) 50 rings (B) 60 rings (C) 70 rings (D) 80 rings

9 The table shows the number of cherries in the crates 11, 12, 13, 14, and 15. Based on the given table, how many cherries are there in crate 14?

Crate	Number of Students
11	110
12	120
13	130
14	?
15	150

(A) 140 (B) 240 (C) 340 (D) 440

10 Sam has a cookie shop. There are 80 different packs of cookies in the shop. Each pack consists of 10 cookies. Complete the following so that the sentence gives the total number of cookies in Sam's Shop.

80 × _____ = (8 × 1) × _____ = _____

(A) 200 (B) 400 (C) 600 (D) 800

49

PLACE VALUE

2.3 Multiply and Divide by Multiples of 10

11. It takes 5 hours to prepare food for 5,000 people. At this rate, how long does it take to prepare food for 15,000 people?

    A) 10 hours   B) 15 hours   C) 20 hours   D) 25 hours

12.
    A) 20×2×4   B) 10×70   C) 4×10×7   D) 4×5×7

13. Justin can pack 100 cakes per hour. How many minutes will it take to pack 350 cakes?

    A) 210 min   B) 230 min   C) 250 min   D) 270 min

14. Lucy's shop has 3,000 pears. She will pack 30 pears in each box. She will sell each box for $20. Determine the amount of money Lucy could make if she sells all boxes of pears.

    A) $1,000   B) $2,000   C) $3,000   D) $4,000

15. There are 50 school buses at the school. Each bus contains 20 seats. How many seats are there in all 50 school buses?

    A) 100   B) 400   C) 800   D) 1,000

PLACE VALUE

2.4 Chapter Review

1. The place value of 5 in 75,310.

   A) Ones   B) Tens   C) Hundreds   D) Thousands

2. The number 56,284 in words is

   A) Fifty thousand two hundred and eighty
   B) Fifty-six thousands two hundred
   C) Fifty-six thousand two hundred and eighty-four
   D) Fifty-six thousands eight hundred and four

3. Write in expanded form, 165,825 is equal to:

   A) 100,000+60,000+5,000+800+20+5
   B) 10,000+6,000+500+20+5
   C) 160,000+5,800+820+20+5
   D) 5+20+5,000+60,000+100,000

PLACE VALUE

2.4 Chapter Review

4 Twenty-two thousand twenty written in standard form is:

A) 2,220   B) 22,020   C) 20,220   D) 22,222

5

6 The numbers 754, 622, 699, and 701 ordered from smallest to largest are

A) 754, 701, 699, and 622
B) 699, 622, 754, and 701
C) 622, 699, 701, and 754
D) 791, 699, 754, and 622

PLACE VALUE

Chapter Review 2.4

7. Place 987, 547, 12, 875, 387, 798, 610, 748, 652, 88, and 1,100 in descending order.

   A) 1,100, 12, 987, 88, 798, 387, 748, 547, 652, 610, 875
   B) 12, 88, 387, 547, 610, 652, 748, 798, 875, 987, 1,100
   C) 1,100, 875, 748, 610, 387, 12, 987, 798, 652, 387, 12
   D) 1,100, 987, 875, 798, 748, 652, 610, 547, 387, 88, 12

8. Sam is writing a number with the following characteristics:

   - 7 ten thousands
   - 1 thousands
   - 5 hundreds
   - 8 tens
   - 4 ones

   What number is Sam writing?

   A) 70,567
   B) 71,584
   C) 71,432
   D) 70,267

9. What number has 9 ten thousands, 2 fewer thousands than ten thousand, 4 fewer hundred than thousands, 2 more tens than hundred and the same number of ones as ten thousands.

   A) 97,359
   B) 96,345
   C) 97,211
   D) 95,479

PLACE VALUE

2.4 Chapter Review

10 Larry obtained 900 when rounding a number to the nearest hundred. Which of the following could be the number that Larry rounded?

A) 978   B) 845   C) 920   D) 9614

11

A) 91,357   B) 75,319   C) 57,913   D) 79,135

12 Which of the following numbers with an underlined digit is worth $1,000?

A) $98,34<u>1</u>   B) $1<u>5</u>,800   C) $2<u>1</u>6,546   D) $<u>1</u>,578

13 Estimate 26,780 ÷ 195

A) 125   B) 135   C) 145   D) 155

14 Which two numbers give the same result when rounded to the nearest 1,000?

A) 5,987 and 4,798   B) 7,527 and 4,708
C) 6,348 and 5,956   D) 1,356 and 3,234

PLACE VALUE

Chapter Review 2.4

15 Susan cooked 48 crabs. Each crab costs $6.
Give an estimate for the total cost of the crabs.

(A) $288 (B) $278 (C) $268 (D) $258

16 Ann bought marbles in a shop. Her bill for the marbles bought is shown on the table.

Marbles	Prize
Red Marbles	$24
Blue Marbles	$22
Green Marbles	$28
White Marbles	$20

(A) $190 (B) $90 (C) $120 (D) $110

17 n is a number. if you multiply n by 140 you get 98,560. What number is n?

(A) 1024 (B) 504 (C) 1004 (D) 704

18 A candle-making machine can produce 500 candles per hour. How many minutes will it take to produce 5,500 candles?

(A) 810 min (B) 530 min (C) 660 min (D) 970 min

PLACE VALUE

2.4 Chapter Review

19 What is the digit in the thousands place that appears in the number 256,831?

(A) 6   (B) 2   (C) 5   (D) 8

20 Write each number in expanded form:

813,562

Sixty-three thousand nine hundred and eighteen.

(A) 8+10+900+3,000+60,000
(B) 800,000+10,000+3,000+500+60+2
(C) 80,000+6,000+300+20+1
(D) 800,000+60,000+5,000+400+30+2

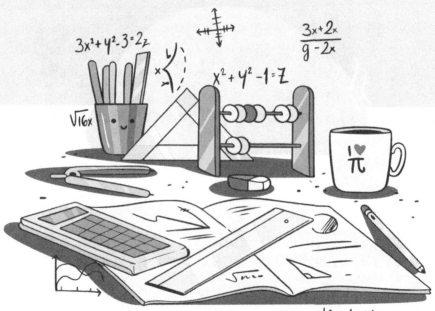

CHAPTER 3
FACTORS AND PATTERNS

FACTORS AND PATTERNS

3.1 Relationship Between Factors and Multiples

Relationship Between Factors and Multiples

The multiples of a natural number n are
$$1\times n, 2\times n, 3\times n, 4\times n, \ldots$$
The multiples of a natural number are the numbers we obtain by multiplying this natural number by 1, 2, 3, ...

Example:

The first 4 multiples of 3 are:
$$1\times 3, 2\times 3, 3\times 3 \text{ and } 4\times 3.$$
They are:
$$3, 6, 9, \text{ and } 12.$$
Multiples of a number can be obtained by skip counting by this number starting from zero.

The first five multiples of 5 are:
$$5, 10, 15, 20, \text{ and } 25.$$
20 is a multiple of 4 $(20\div 4=5)$ while 21 is not.

FACTORS AND PATTERNS

Relationship Between Factors and Multiples — 3.1

1. How many factor pairs does the number 36 have?

   (A) 4   (B) 6   (C) 8   (D) 10

2. Which number is not a factor of 52?

   (A) 1   (B) 52   (C) 8   (D) 26

3. Which number is a factor of 81?

   (A) 13   (B) 17   (C) 23   (D) 27

4. Which number has 9 as one of its factors?

   (A) 121   (B) 162   (C) 197   (D) 221

5. Which number does not have 4 as one of its factors?

   (A) 184   (B) 156   (C) 132   (D) 102

FACTORS AND PATTERNS

3.1 Relationship Between Factors and Multiples

6 What are all the factor pairs of 76? How do you know?

_____

_____

7 Find the first five multiples of 14

A) _____ D) _____

B) _____ E) _____

C) _____

8 A number is a multiple of 16. Which of the following must be a factor in this number?

A) 3   B) 6   C) 5   D) 4

9 Which number is a factor of 33 but not a multiple of 3?

A) 2   B) 3   C) 11   D) 33

FACTORS AND PATTERNS

Relationship Between Factors and Multiples — 3.1

10 Use this information to show why 60 must also be a multiple of 2.

- A) 2×13=26 and 26×2=60
- B) 2×15=30 and 30×2=60
- C) 2×17=34 and 34×2=60
- D) 2×19=38 and 38×2=60

11 Among the numbers 45, 30, 27, 18, 21, 42, and 72, choose the numbers that are divisors of the number 9 and multiples of the number 3 simultaneously.

- A) 18, 27, 45, and 72
- B) 72, 45, 30, 27, and 18
- C) 21, 27, 30, and 42
- D) 18, 21, 42, and 72

12 Mark ordered 60 boxes of pencils. Each box contains 4 packs of pencils, and each pack contains 15 pencils. How are these three numbers related?

- A) 60 is a multiple of 4 and 15, 15 and 4 are divisors of the number 60.
- B) 60 is a multiple of 6 and 8, 8 and 6 are divisors of the number 60.

FACTORS AND PATTERNS

3.1 Relationship Between Factors and Multiples

C) 60 is a multiple of 4 and 10, 10 and 4 are divisors of the number 60.

D) None of these

13. Jin has 42 color pens in a pack.

    If he arranges the color pens in 6 rows, there will be _____ color pens in each row. If he arranges the color pens in _____ rows, there will be 7 color pens in each row.

14. Use the numbers 21 and 7 to complete the following sentences.

    _____ is a factor of _____

    _____ is a multiple of _____

15. A number that has exactly two factors is called a _____ number.

NEXT CHAPTER:
3.2 Predict and Extend Growing and Repeating Patterns

FACTORS AND PATTERNS

3.2 Predict and Extend Growing and Repeating Patterns

Predict and Extend Growing and Repeating Patterns

A number pattern is a sequence or list of numbers that are related. These numbers are generated through a rule that may be deduced by studying the pattern.

Number patterns may be observed through objects that we encounter in our daily life. For example, knowing that the height of one story in a building is 3 m, the levels of the floors in that building starting from the first floor form the sequence 3, 6, 9, 12, etc. This is a pattern generated when we count by 3, or equivalently add 3.

The elements of the pattern are called terms.

The first term in the above pattern is 3, the second term is 6, and so on.

Example: Fill in the missing entries in table 1.

Input-output rule: $3n - 1$

Table 1

Input	1	3	4	10	11
Output					

Solution

Input	1	3	4	10	11
Output	2	8	11	29	32

FACTORS AND PATTERNS

3.2 Predict and Extend Growing and Repeating Patterns

1. Enter the correct number in the pattern 2, 9, 16, 23, _____ .

   A) 30   B) 31   C) 32   D) 33

2. Enter the correct number in the pattern. 17, 51, 153, 459, _____, 4,131

   A) 2,023   B) 1,791   C) 1,377   D) 1,111

3. The first number is 14. The rule is to multiply by 4 and add 4. Determine the sequence by listing the numbers below.

   A) _____   D) _____

   B) _____   E) _____

   C) _____

FACTORS AND PATTERNS

Predict and Extend Growing and Repeating Patterns 3.2

4 The rule adds 5 and multiplying by 2 is applied to each number in a column labeled Number.

Number	Number add 5 Multiply 2
2	14
4	18
6	22
8	26

If the same rule is applied to the number 22, the result would be

A) 54   B) 58   C) 62   D) 66

5 The rule to multiply by 5 and subtract 2 is applied to the numbers 2, 4, 6, and 8 in this order. What can we conclude about the resulting numbers?

A) They are all odd

B) They are all even

C) The even-odd alternate, with the first number being even

D) The alternate odd-even pattern has the first number being odd

6 Amy creates patterns 4, 7, 12, 15, 20, and 23. What are the next five numbers in Amy's pattern?

A) 27, 32, 35, 38, and 43

B) 25, 30, 33, 39, and 42

C) 26, 31, 34, 37, and 42

D) 28, 31, 36, 39, and 44

FACTORS AND PATTERNS

3.2 Predict and Extend Growing and Repeating Patterns

7 A pattern consists of odd numbers and has a rule of adding 6. Which set of numbers matches this pattern description?

- (A) 0, 2, 4, 6, and 8
- (B) 3, 9, 15, 21, and 27
- (C) 1, 7, 11, 17, and 19
- (D) 6, 12, 18, 24, and 30

8 The first two terms of a pattern are 1 and 3. The third term is obtained by multiplying the previous two terms. Find the first six terms of this pattern.

- (A) 1, 3, 3, 9, 27, and 243
- (B) 1, 3, 9, 27, 81, and 243
- (C) 1, 3, 5, 7, 9, and 11
- (D) 1, 3, 3, 12, 24, and 124

9 Nick makes muffins in his shop on a regular basis, and each time he can increase the number of muffins. After reviewing the production rate, Nick saw that every week for a month, he made a maximum of 352 at first, then 371, then 390, then 409 muffins. What will be his record for muffins next week?

- (A) 437
- (B) 438
- (C) 428
- (D) 427

10 After adding 25 different species of animals to the zoo, the number of visitors increased. On Sunday 140 visitors, on Monday 165 visitors, on Tuesday 190 visitors, and on Wednesday 215 visitors. How many visitors will visit the zoo on Thursday?

- (A) 235
- (B) 240
- (C) 245
- (D) 250

FACTORS AND PATTERNS

Predict and Extend Growing and Repeating Patterns — 3.2

11. The number pattern below is obtained by multiplying by 5 and 6 alternatively.

1, 5, 30, 150, 900, _____ , _____

What is the next term in the pattern?

- (A) 1,500 and 9,000
- (B) 2,500 and 15,000
- (C) 3,500 and 21,000
- (D) 4,500 and 27,000

12. Consider the arithmetic pattern

7, _____, 41, 58, 75, 92

What number should be used to fill in the blank?

- (A) 24
- (B) 34
- (C) 22
- (D) 32

13. If the number pattern below is a geometric progression, what number must go in the blank?

2, _____, 32, 128, 512, 2,048

- (A) 18
- (B) 24
- (C) 8
- (D) 12

FACTORS AND PATTERNS

3.2 Predict and Extend Growing and Repeating Patterns

14 The table shows the number of butterflies in the garden on different days.

Day	Number of butterflies
1	125
2	140
3	155
4	?

How many butterflies are there on day 4?

A) 160   B) 170   C) 180   D) 190

15 Tom solves math problems every day. He needs to be faster to solve all the problems per day. On day 1 he solved 12 problems, on day 2 he solved 23 problems, and on day 3 he solved 34 problems. How many problems did Tom solve on day 4?

A) 45   B) 43   C) 41   D) 47

NEXT CHAPTER:
3.3 Pattern Rules

FACTORS AND PATTERNS

3.3 Pattern Rules

Pattern Rules

Just about any mathematical rule can be used to define a number pattern. One must be careful when identifying a rule that defines a given pattern. Every number listed in the pattern must be tested as to whether it is generated by this rule or not.

Tables of values

Given the table of corresponding values for two unknowns, it is possible to find the rule that generates this table. We first observe the pattern of changes in the values of the unknowns, and express the relation as a word sentence, then translate it into a number sentence. This is illustrated in the following example.

Input-output table

An input-output table is a two-column (or two-row) table. One column (or row) is for the input, while the other is for the output. Each table has a rule. When a number n is put in for the input, it is changed by the rule to give a different number for the output.

Example:

Using the following table of values, find a rule that describes the relation between:

Number (x)	1	2	3	4	5
Rule (y)	5	9	13	17	21

FACTORS AND PATTERNS

3.3 Pattern Rules

Pattern Rules

Solution:

Note that each unit increment in the value of "Number" corresponds to an increase of 4 in the value of "Rule." When "number" is equal to 1, the corresponding value of "rule" is obtained by multiplying 1 by 4, and then we have to add 1 to get 5. We conclude the following:

$$5 = 4 \times 1 + 1 = 4 + 1$$
$$9 = 4 \times 2 + 1 = 8 + 1$$
$$13 = 4 \times 3 + 1 = 12 + 1$$
$$17 = 4 \times 4 + 1 = 16 + 1$$
$$21 = 4 \times 5 + 1 = 20 + 1$$

It follows that the rule for generating the table is to multiply by 4 and add 1. As a number sentence we write, $y = 4x + 1$.

FACTORS AND PATTERNS

Pattern Rules — 3.3

1 Rose travels 70 kilometers an hour. What number is missing in the table?

Hours	Kilometers
1	70
7	490
13	?
15	1,050

(A) 930   (B) 890   (C) 910   (D) 860

2 What is the next number in the series 5, 20, 80, 320?

(A) 640   (B) 1,280   (C) 960   (D) 1,420

3 What is true about a number pattern that starts at 270 with a rule of "add 3"?

(A) Numbers in the pattern will be even.

(B) Numbers in the pattern will be odd.

(C) Numbers in the pattern will alternate between even and odd.

(D) Numbers in the pattern will be even twice, then odd once.

FACTORS AND PATTERNS

3.3 Pattern Rules

4. What is true about a number pattern that starts at 500 with a rule of "subtract 2"?

   A) Numbers in the pattern will be even.
   B) Numbers in the pattern will be odd.
   C) Numbers in the pattern will alternate between even and odd.
   D) Numbers in the pattern will be even twice, then odd once.

5. What is the next number in this pattern?
   8, 16, 24, 32.....

   A) 36   B) 40   C) 42   D) 48

6. What number goes in the blank to complete the pattern?
   51, _____, 63, 69, 75

   A) 55   B) 59   C) 53   D) 57

7. What number goes in the blank to complete the pattern?
   98, _____, 88, 83, 78.

8. What is the rule of the sequence 71, 87, 103,...?
   The rule is _____.

FACTORS AND PATTERNS

Pattern Rules — 3.3

9 The first number in the sequence is 28. The rule is "add 12." What is the fifth number in the sequence?

10 Part of a sequence is ... 93, 102, 111...
What number goes directly before 93 in the sequence?

11 What number is missing from this sequence?
...... 270, _____ , 30, 10

12 The first number in a sequence is 200. The rule is "divided by 2 then add 10."
What number will be third in the sequence?

FACTORS AND PATTERNS

3.3 Pattern Rules

13 The fourth number in a sequence is 107. The rule is "add 15." What is the first number in the sequence?

14 If Serena starts with 10 and makes a pattern with the rule "add 8," will the numbers in her pattern be even, odd, or a combination of both? How do you know?

15 What rule is used to create this pattern? Explain how you know. 6, 12, 18, 24, 30.

NEXT CHAPTER:
3.4 Prime and Composite Numbers

FACTORS AND PATTERNS

3.4 Prime and Composite Numbers

Prime and Composite Numbers

A prime number is a number that has two and only two different factors. The factors are the number itself and 1.

The way of expressing a number as the product of its prime factors is called the prime factorization of that number.

A number that has more than two factors is called a composite number. 0 and 1 are neither prime nor composite numbers.

Example:

1. List the natural numbers from 1 to 10, underline the prime numbers in your list, and circle the composite ones.
2. How many numbers in your list are neither prime nor composite?

Solution:

(A) 1   <u>2</u>   <u>3</u>   4   <u>5</u>

    6   <u>7</u>   8   9   10

(B) There is one number in the list that is neither prime nor composite.

Prime Factorization

The way of expressing a number as the product of its prime factors is called the prime factorization of that number.

FACTORS AND PATTERNS

3.4 Prime and Composite Numbers

1. Which of the following numbers is not a composite number?
   - A) 53
   - B) 51
   - C) 55
   - D) 49

2. Which of the following numbers is not prime?
   - A) 37
   - B) 61
   - C) 97
   - D) 108

3. What is the next prime number after 31?
   - A) 33
   - B) 35
   - C) 37
   - D) 39

4. Is 73 a prime or a composite number? How do you know?

   _____

   _____

5. What is the next prime number after 41? How do you know?

   _____

   _____

FACTORS AND PATTERNS

Prime and Composite Numbers — 3.4

6. A number that has more than one factor pair is called a _____ number.

7. How many prime numbers are between 11 and 30?

8. What is the next composite number after 70?

9. How many prime numbers are there between 101 and 150?

FACTORS AND PATTERNS

3.4 Prime and Composite Numbers

10 Is the number 101 prime or composite?

_____

_____

11 Which of the following numbers is a composite number?

- (A) 37
- (B) 100
- (C) 109
- (D) 137

12 Rosie wrote the number pattern: 3, 7, 15, 31, and 63. Which of the following rules can be used to find the next number in the number pattern?

- (A) Add 4
- (B) Multiply by 3 and subtract 2
- (C) Multiply by 2 and add 1
- (D) Subtract 3

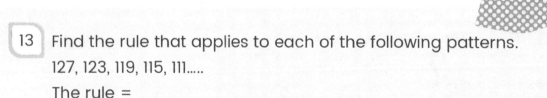

13 Find the rule that applies to each of the following patterns.
127, 123, 119, 115, 111…..
The rule = _____

FACTORS AND PATTERNS

Prime and Composite Numbers — 3.4

14 Based on the pattern in the table below, what is the input-output rule?

Input	2	4	6	8	10
Output	6	7	8	9	10

15 The teacher knows that there is a relationship between the time needed to solve the problems and the number of exercises. He compiled a table to understand this pattern.

Number of exercises	Time to solve problems (hours)
3	22
6	16
9	14
12	13

Find out the time needed?

A) 36 hours divided by the number of exercises, and add 10 hours.

B) 42 hours divided by the number of exercises, and add 8 hours.

C) 64 hours divided by the number of exercises, and add 6 hours.

D) 72 hours divided by the number of exercises, and add 10 hours.

FACTORS AND PATTERNS

3.5 Chapter Review

1. How many factor pairs does 120 have?

   A) 12   B) 10   C) 8   D) 6

2. The number 5 and what other numbers make a factor pair of 65?

   A) 12   B) 13   C) 14   D) 15

3. Is the number 98 prime or composite?

   _____

   _____

4. What is the next number in the pattern? 223, 244, 265, 286 .....

   A) 307   B) 314   C) 326   D) 334

5. What is the next number in the pattern? 4,500, 900, 180 .....

   A) 34   B) 24   C) 56   D) 36

FACTORS AND PATTERNS

Chapter Review 3.5

6 Peter wrote the following pattern: 19, 38, 76, ...
What is the rule of his pattern?

- A) Add 19
- B) Multiply by 2
- C) Add 21
- D) Multiply by 3

7 Part of a sequence is ...792, 819, 846, 873 ...
What number comes directly before 792 in the sequence?

- A) 746
- B) 759
- C) 765
- D) 783

8 Roy is thinking of a number that is 16 less than 81 divided by 3. What number is Roy thinking of?

9 Daniel bought 68 pens from Mary's shop and 52 pens from James' shop. He is going to put the pens into boxes of 8. How many full boxes will Daniel be able to make?

FACTORS AND PATTERNS

3.5　Chapter Review

10 A rule is applied to each number in the column labeled "Number of Calculators" to produce the corresponding number in the second column labeled "Dollars".

Number of Calculators	Dollars
5	25
10	35
15	45
20	55
25	65

What rule is being applied?

A) Multiply by 5

B) Add 15

C) Multiply 2 and then add 15

D) Multiply 5 and then subtract 15

11 Joseph sold 2,750 cups in boxes that could fit 25 cups each. How many boxes did Joseph use?

A) 110   B) 105   C) 115   D) 120

12 Mr. Michael wants to split the cars in his showroom into 3 groups with exactly the same number of cars in each group. He later realized that this was not possible. Which of the following could be the number of cars in Mr. Michael's showroom?

A) 534   B) 516   C) 525   D) 503

FACTORS AND PATTERNS

Chapter Review 3.5

13. Tom bought 42 shirts from John's shop and 30 shirts from Jerry's shop. He is going to put the shirts into boxes of 6. How many full boxes will Tom be able to make?

    (A) 10   (B) 12   (C) 14   (D) 16

14. Which list contains all composite numbers?

    (A) 53, 23, 11, and 71
    (B) 19, 29, 37, and 47
    (C) 48, 51, 62, and 81
    (D) 59, 67, 71, and 73

15. Which list contains two prime numbers?

    (A) 11, 18, 27, 34, 52, and 62
    (B) 22, 34, 41, 55, 57, and 81
    (C) 30, 45, 69, 71, 84, and 90
    (D) 28, 36, 47, 51, 53, and 63

16. Which of the following is a multiple of both 8 and 13?

    (A) 96   (B) 104   (C) 78   (D) 117

FACTORS AND PATTERNS

3.5 Chapter Review

17 Which of the following is not a factor pair of 120?

(A) 4 & 40   (B) 4 & 30   (C) 2 & 60   (D) 10 & 12

18 I am a prime number with two digits. Each of my digits is prime, and the difference between them is 4.

(A) 19   (B) 73   (C) 59   (D) 47

19 Flower pots are arranged in rows. There are 250 flower pots in each row. In all, there are 12,000 flower pots. What is the number of rows?

(A) 72   (B) 64   (C) 56   (D) 48

20 If you multiply Nora's age by 500 and add 700, you get 5,700. How old is Nora?

(A) 8   (B) 10   (C) 12   (D) 14

CHAPTER 4

FRACTIONS

FRACTIONS

4.1 Equivalent Fraction Using an Area Model

Equivalent Fraction Using an Area Model

A fraction such as $\dfrac{3}{8}$ represents a division problem; the top number (3), called the **numerator** of the fraction, is divided by the bottom number (8), called the **denominator** of the fraction.

Equivalent fractions

Fractions that represent the same amount are called equivalent fractions.

$$\dfrac{2}{4} = \dfrac{1}{2}$$

We can use fraction strips to help you find equivalent fractions.

Example: Two-fourths are equivalent to one-half.

So, $\dfrac{2}{4} = \dfrac{1}{2}$

Two-fourths are also equivalent to four-eighths.

So, $\dfrac{2}{4} = \dfrac{4}{8}$

FRACTIONS

Equivalent Fraction Using an Area Model 4.1

1. Which fraction is equivalent to $\frac{6}{30}$?

   (A) $\frac{1}{10}$ (B) $\frac{6}{10}$ (C) $\frac{60}{300}$ (D) $\frac{6}{10}$

2. Isabella has 40 straws. She has 24 pink straws and 16 violet straws. Which fraction is equivalent to the fraction ratio of the violet straws to the total straws?

   (A) $\frac{2}{5}$ (B) $\frac{4}{5}$ (C) $\frac{3}{5}$ (D) $\frac{6}{5}$

3. The shaded part of the model below represents a fraction of the total area of the model.

   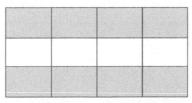

   Which of the following fractions is equivalent to the shaded portion in the model?

   (A) $\frac{2}{4}, \frac{16}{20}, \frac{24}{30}$ (B) $\frac{2}{3}, \frac{16}{24}, \frac{24}{36}$ (C) $\frac{4}{3}, \frac{16}{25}, \frac{24}{35}$ (D) $\frac{2}{3}, \frac{26}{40}, \frac{24}{60}$

4. Which fraction has the greatest value?

   (A) $\frac{7}{19}$ (B) $\frac{5}{3}$ (C) $\frac{14}{12}$ (D) $\frac{9}{4}$

FRACTIONS

4.1 Equivalent Fraction Using an Area Model

5 Order the fractions from greatest to smallest?

$$\frac{4}{10}, \frac{70}{100}, \frac{2}{25}, \frac{9}{20}$$

_____

_____

6 What value must replace the box in $\frac{4}{10} = \frac{?}{12}$

$\frac{1}{12}$	$\frac{1}{12}$	$\frac{1}{12}$	$\frac{1}{12}$	$\frac{1}{12}$	$\frac{1}{12}$	$\frac{1}{12}$	$\frac{1}{12}$	$\frac{1}{12}$	$\frac{1}{12}$	$\frac{1}{12}$	$\frac{1}{12}$
$\frac{1}{4}$			$\frac{1}{4}$			$\frac{1}{4}$			$\frac{1}{4}$		

(A) $\frac{3}{6}$ (B) $\frac{9}{12}$ (C) $\frac{9}{10}$ (D) $\frac{4}{8}$

7 Use the diagram to complete the equivalent fraction sentence

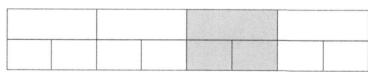

$$\frac{1}{2} = \frac{?}{10}$$

(A) $\frac{4}{10}$ (B) $\frac{3}{8}$ (C) $\frac{5}{10}$ (D) $\frac{2}{10}$

FRACTIONS

Equivalent Fraction Using an Area Model — 4.1

8 Use the models to complete the equivalent fraction sentence.

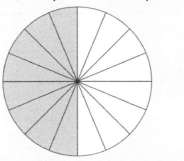 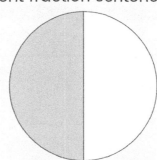

$$\frac{?}{16} = \frac{1}{2}$$

(A) $\frac{8}{16}$ (B) $\frac{10}{16}$ (C) $\frac{12}{16}$ (D) $\frac{4}{16}$

9 Use the models to complete the equivalent fraction sentence.

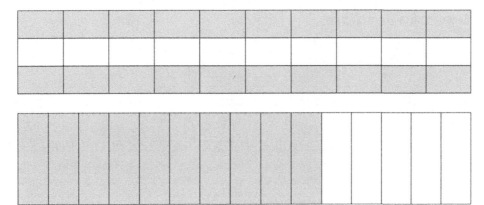

$$\frac{10}{?} = \frac{20}{30}$$

(A) $\frac{10}{30}$ (B) $\frac{10}{20}$ (C) $\frac{10}{5}$ (D) $\frac{10}{15}$

FRACTIONS

4.1 Equivalent Fraction Using an Area Model

10 Give two equivalent fractions that the diagram below suggests.

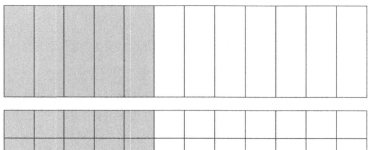

Ⓐ $\dfrac{10}{12} = \dfrac{18}{36}$ Ⓑ $\dfrac{8}{10} = \dfrac{15}{30}$ Ⓒ $\dfrac{5}{12} = \dfrac{15}{36}$ Ⓓ $\dfrac{15}{20} = \dfrac{15}{30}$

11 Bella bought one cake. The shaded part of the diagram below represents the fraction of the cake Bella ate.

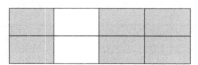

Bella ate $\dfrac{3}{4}$ of the cake. $\dfrac{3}{4}$ is equivalent to $\dfrac{k}{8}$. What is the value of k?

_____

_____

FRACTIONS

Equivalent Fraction Using an Area Model — 4.1

12 Emily and Ella bought two apples.
- Emily cut her apple into 4 equal slices and ate 3 of them
- Ella cut her apple into 8 equal slices and ate a number of slices.

If the two girls ate the same amount of apples, how many apple slices did Ella eat?

A) 10   B) 8   C) 4   D) 6

13 James had pie for dinner. The shaded area in the diagram below represents the fraction of the pie that James ate. What fraction of the pie did James eat?

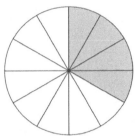

A) $\dfrac{10}{30}$ B) $\dfrac{10}{20}$ C) $\dfrac{10}{5}$ D) $\dfrac{10}{15}$

14 Use the diagram below to answer the question.

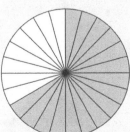

Ann ate $\dfrac{16}{24}$ of a pizza. $\dfrac{16}{24}$ is equivalent to $\dfrac{k}{12}$. What is the value of k?

A) 10   B) 6   C) 8   D) 12

FRACTIONS

4.1 Equivalent Fraction Using an Area Model

15. Write the equivalent fractions shown by the shaded area in the pair of diagrams.

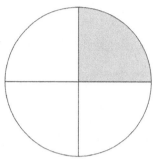 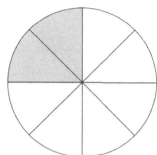

Ⓐ $\dfrac{17}{19} = \dfrac{4}{8}$ Ⓑ $\dfrac{1}{4} = \dfrac{2}{8}$ Ⓒ $\dfrac{3}{4} = \dfrac{2}{8}$ Ⓓ $\dfrac{9}{9} = \dfrac{14}{14}$

QUICK TIPS

Whole	One Half	One Third	One Quarter	One Sixth	One Eighth

$\dfrac{1}{1}$	$\dfrac{1}{2}$	$\dfrac{1}{3}$	$\dfrac{1}{4}$	$\dfrac{1}{6}$	$\dfrac{1}{8}$

NEXT CHAPTER:
4.2 Equivalent Fraction Using a Length Fraction Model

FRACTIONS

4.2 Equivalent Fraction Using a Length Fraction Model

Notes – Equivalent Fraction Using a Length Fraction Model

Two fractions are equivalent if they represent the same amount. Multiplying the numerator and the denominator of a fraction by the same non-zero number produces an equivalent fraction.

For instance, $\frac{1}{3} = \frac{1 \times 2}{3 \times 2} = \frac{1 \times 3}{3 \times 3} = \frac{1 \times 4}{3 \times 4}$, simplifying one obtains $\frac{1}{3} = \frac{2}{6} = \frac{3}{9} = \frac{4}{12}$ and this concludes that $\frac{1}{3}$, $\frac{2}{6}$, $\frac{3}{9}$ and $\frac{4}{12}$ are equivalent fractions.

Using the number line drawn below, one may conclude that $\frac{1}{3}$, $\frac{2}{6}$, $\frac{3}{9}$ and $\frac{4}{12}$ are equivalent fractions.

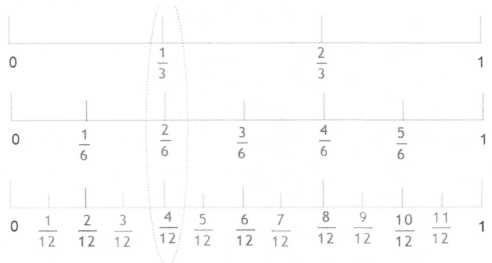

Dividing the numerator and the denominator of a fraction by a common factor is called canceling, and it also produces an equivalent fraction. This fact enables us to represent fractions using smaller numbers, which we refer to as the simplification of a fraction.

FRACTIONS

4.2 Equivalent Fraction Using a Length Fraction Model

1. Complete the following:

$$\frac{5}{9} = \frac{5 \times \ldots}{9 \times 2} = \frac{5 \times 3}{9 \times \ldots} = \frac{5 \times \ldots}{9 \times 4} \text{ therefore, } \frac{5}{9} = \frac{\square}{\square} = \frac{\square}{\square} = \frac{\square}{\square}$$

(A) $\dfrac{5}{9} = \dfrac{10}{18} = \dfrac{15}{27} = \dfrac{20}{36}$ (B) $\dfrac{5}{9} = \dfrac{9}{5} = \dfrac{10}{18} = \dfrac{18}{10}$

(C) $\dfrac{20}{36} = \dfrac{10}{18} = \dfrac{5}{9} = \dfrac{9}{5}$ (D) $\dfrac{5}{10} = \dfrac{11}{15} = \dfrac{17}{20} = \dfrac{21}{15}$

2. Complete the following:

$$\frac{26}{52} = \frac{26 \div 2}{52 \div 2} = \frac{\ldots \div 13}{\ldots \div 13} \text{ therefore, } \frac{26}{52} = \frac{\square}{\square} = \frac{\square}{\square}$$

(A) $\dfrac{26}{51} = \dfrac{13}{39} = \dfrac{1}{3}$ (B) $\dfrac{13}{21} = \dfrac{15}{31} = \dfrac{21}{37}$

(C) $\dfrac{26}{52} = \dfrac{13}{26} = \dfrac{2}{4}$ (D) $\dfrac{26}{52} = \dfrac{2}{6} = \dfrac{1}{13}$

3. Graph $\dfrac{5}{6}$ on the number line:

FRACTIONS

Equivalent Fraction Using a Length Fraction Model   4.2

4 Graph $\frac{70}{100}$ on the number line:

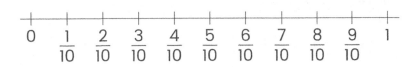

5 In Grade 4, 36 out of 54 students are girls. Express, in the lowest terms, the fraction of the students in the class that are girls.

- A) $\frac{4}{3}$
- B) $\frac{7}{3}$
- C) $\frac{2}{3}$
- D) $\frac{5}{3}$

6 Mary bought 120 vegetables, 20 of which are potatoes. How can we express the number of potatoes as a fraction?

- A) $\frac{1}{4}$
- B) $\frac{1}{6}$
- C) $\frac{1}{5}$
- D) $\frac{1}{3}$

7 Austin gets $\frac{15}{9}$ of apples. Fill in the missing number in the box below to show how many thirds of apple must get so that he gets the same amount of apples as Austin: $\frac{15}{9} = \frac{\boxed{}}{3}$

- A) 5
- B) 3
- C) 7
- D) 1

FRACTIONS

4.2 Equivalent Fraction Using a Length Fraction Model

8 There are 48 ribbons on the table. 12 of the ribbons are black. What fraction of the ribbons are black?

(A) $\frac{9}{13}$ (B) $\frac{7}{4}$ (C) $\frac{13}{8}$ (D) $\frac{1}{4}$

9 There were 56 erasers inside a box. 7 erasers are broken. That is $\frac{7}{56}$ of all erasers are broken. What fraction of the erasers are broken inside the box that is equivalent to $\frac{7}{56}$?

(A) $\frac{1}{7}$ (B) $\frac{8}{52}$ (C) $\frac{1}{8}$ (D) $\frac{7}{16}$

10 Five of 45 equal pieces is the same as _____ of 9 equal pieces.

11 What fraction is equivalent to $\frac{4}{7}$ and has a denominator of 21?

(A) $\frac{8}{21}$ (B) $\frac{12}{21}$ (C) $\frac{16}{21}$ (D) $\frac{4}{21}$

12 What value ☐ would make the equation $\frac{\square}{48} = \frac{2}{3}$ true?

(A) 42 (B) 28 (C) 32 (D) 38

FRACTIONS

Equivalent Fraction Using a Length Fraction Model — 4.2

13 Out of 180 marbles in a shop, 65 marbles are in red color, 80 marbles are in blue color, and 35 marbles are in green color. Express, in the lowest terms, the fraction of the 180 marbles that are in blue color?

___

___

14 Anya, Clara, and Elisa are friends. They chose a fraction between 1 and 2. Anya chose $\frac{4}{3}$, Clara chose $\frac{7}{6}$ and Elisa chose $\frac{12}{10}$. Select the group that includes an equivalent fraction for each of the fractions $\frac{4}{3}$, $\frac{7}{6}$, and $\frac{12}{10}$.

(A) $\frac{4}{3}, \frac{7}{6}$ and $\frac{12}{10}$ (B) $\frac{5}{3}, \frac{7}{9}$ and $\frac{2}{10}$

(C) $\frac{4}{8}, \frac{12}{6}$ and $\frac{4}{10}$ (D) $\frac{6}{12}, \frac{21}{28}$ and $\frac{6}{5}$

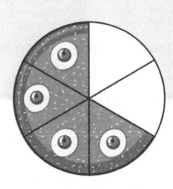

15 Give two fractions that are equivalent to $\frac{17}{51}$

___

___

NEXT CHAPTER:
4.3 Add and Subtract Fraction

FRACTIONS

4.3 Add and Subtract Fractions

Add and Subtract Fractions

Eight parts $\frac{8}{9}$ in two separate regions of the rectangle are shaded.

Five equal parts in one region $\frac{5}{9}$ and three parts in the other $\frac{3}{9}$.

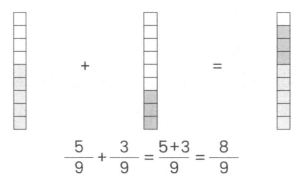

$$\frac{5}{9} + \frac{3}{9} = \frac{5+3}{9} = \frac{8}{9}$$

This fact gives the rule for adding two fractions with like denominators:

Add the numerators.

Subtracting fractions with like denominators is done the same way:

Subtract the numerators.

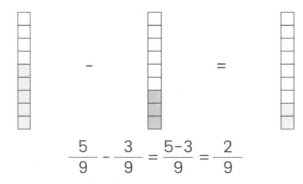

$$\frac{5}{9} - \frac{3}{9} = \frac{5-3}{9} = \frac{2}{9}$$

To add or subtract fractions with denominators, simply add or subtract the numerators and express the result as a fraction whose denominator is the common denominator.

FRACTIONS

Add and Subtract Fraction 4.3

1 Which addition expression represents the total shaded parts in this fraction model?

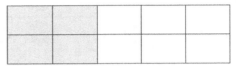

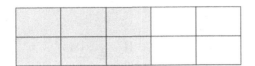

Ⓐ $\dfrac{2}{10} + \dfrac{3}{10}$ Ⓑ $\dfrac{4}{10} + \dfrac{6}{10}$ Ⓒ $\dfrac{5}{10} + \dfrac{7}{10}$ Ⓓ $\dfrac{8}{10} + \dfrac{9}{10}$

2 Which expression has a sum that equals $\dfrac{5}{7}$?

Ⓐ $\dfrac{1}{7} + \dfrac{2}{7} + \dfrac{3}{7}$ Ⓑ $\dfrac{2}{7} + \dfrac{4}{7} + \dfrac{3}{7} + \dfrac{5}{7}$

Ⓒ $\dfrac{1}{7} + \dfrac{2}{7} + \dfrac{3}{7} + \dfrac{4}{7}$ Ⓓ $\dfrac{1}{7} + \dfrac{1}{7} + \dfrac{1}{7} + \dfrac{1}{7} + \dfrac{1}{7}$

3 Find the sum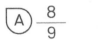

Ⓐ $\dfrac{8}{9}$ Ⓑ $\dfrac{4}{18}$ Ⓒ $\dfrac{7}{27}$ Ⓓ $\dfrac{6}{9}$

FRACTIONS

4.3 Add and Subtract Fraction

4 Find the difference $\dfrac{10}{12} - \dfrac{7}{12}$

- Ⓐ $\dfrac{17}{12}$
- Ⓑ $\dfrac{7}{6}$
- Ⓒ $\dfrac{3}{12}$
- Ⓓ $\dfrac{1}{6}$

5 Which sum is greater than 3?

- Ⓐ $\dfrac{3}{7} + \dfrac{5}{7} = \dfrac{8}{7}$
- Ⓑ $\dfrac{7}{9} + \dfrac{11}{9} = \dfrac{18}{9}$
- Ⓒ $\dfrac{4}{8} + \dfrac{6}{8} = \dfrac{10}{8}$
- Ⓓ $\dfrac{4}{3} + \dfrac{8}{3} = \dfrac{12}{3}$

6 Emma ate $\dfrac{8}{12}$ of a whole watermelon. How much of the watermelon left?

- Ⓐ $\dfrac{6}{12}$
- Ⓑ $\dfrac{4}{12}$
- Ⓒ $\dfrac{10}{12}$
- Ⓓ $\dfrac{2}{12}$

7 Olivia has black, white, and grey T-shirts
- Two-sevenths of the T-shirts are black.
- Four-sevenths of the T-shirts are white.
- There are 49 T-shirts in total.

Determine the number of black T-shirts that can be used.

- Ⓐ 14
- Ⓑ 10
- Ⓒ 8
- Ⓓ 12

FRACTIONS

Add and Subtract Fraction 4.3

8. John bought 20 lemons. He used $\frac{4}{5}$ of the lemons for the juice. How many lemons did he use for the juice?

A) 18   B) 14   C) 16   D) 17

9. Add and express your answer in the simplest form. $\frac{3}{13} + \frac{5}{13} + \frac{8}{13}$

A) $\frac{8}{13}$ B) $\frac{16}{13}$ C) $\frac{11}{13}$ D) $\frac{14}{13}$

10. What sign makes the statement true? $\frac{12}{30} + \frac{16}{30} \square \frac{28}{30}$

A) <   B) >   C) =   D) None of these

11. $\frac{10}{80}$ and $\frac{k}{40}$ are equivalent fractions. Find the value of k

A) 5   B) 6   C) 9   D) 10

12. Max mixed $\frac{3}{13}$ cups of carrot juice and $\frac{7}{13}$ cups of beetroot juice. What is the total number of cups in the juice mix.

A) $\frac{1}{13}$ B) $\frac{10}{13}$ C) $\frac{7}{13}$ D) $\frac{3}{13}$

FRACTIONS

4.3 Add and Subtract Fraction

13. Kelly is mixing different colors. $\frac{6}{12}$ is the red color, $\frac{5}{12}$ is the white color, and the rest is pink. What part of Kelly's mix is pink?

  Ⓐ $\frac{2}{12}$ Ⓑ $\frac{9}{12}$ Ⓒ $\frac{1}{12}$ Ⓓ $\frac{7}{12}$

$\frac{1}{8} + \frac{3}{4} = \dfrac{\square}{\square}$

$\frac{3}{8} + \frac{5}{8} = \dfrac{\square}{\square}$

$\frac{1}{2} + \frac{1}{4} = \dfrac{\square}{\square}$

FRACTIONS

Add and Subtract Fraction — 4.3

14 On Monday, William cut $\frac{7}{10}$ of the vegetables; on Tuesday, he cut $\frac{77}{100}$ of the vegetables. What fraction of the vegetables did William cut on both days?

Ⓐ $\frac{124}{100}$ Ⓑ $\frac{77}{10}$ Ⓒ $\frac{47}{10}$ Ⓓ $\frac{147}{100}$

15 Given $t + \frac{17}{29} = \frac{22}{29}$. Find t.

Ⓐ $\frac{12}{29}$ Ⓑ $\frac{5}{29}$ Ⓒ $\frac{18}{29}$ Ⓓ $\frac{21}{29}$

NEXT CHAPTER:
4.4 Mixed Numbers

FRACTIONS

4.4 Mixed Numbers

Mixed Numbers

Numbers that contain whole as well as well as fractional parts are called mixed numbers.

There is a shaded circle and $\frac{2}{4}$ of a shaded circle.

This number is written as $1\frac{2}{4}$

There are 3 shaded circle and $\frac{1}{4}$ of a shaded circle.

This number is written as $3\frac{1}{4}$

The numbers $1\frac{2}{4}$ and $3\frac{1}{4}$ are called **mixed numbers**, and the value of each is greater than 1.

Add. Remember to add whole numbers to whole numbers and fractions to fractions.

$$1\frac{2}{4} + 3\frac{1}{4} = 1 + 3 + \frac{2}{4} + \frac{1}{4} = 4 + \frac{3}{4} = 4\frac{3}{4}$$

FRACTIONS

Mixed Numbers 4.4

1 Choose the expression that shows adding $5\frac{3}{2} + 4\frac{3}{2}$ using equivalent fractions.

(A) $\frac{13}{2} + \frac{11}{1}$ (B) $\frac{10}{2} + \frac{8}{2}$ (C) $\frac{15}{2} + \frac{12}{2}$ (D) $\frac{20}{2} + \frac{9}{2}$

2 This week, Brody spent $4\frac{5}{3}$ hours playing on Saturday and $3\frac{7}{3}$ hours playing on Sunday. How many groups of $\frac{1}{3}$ hours did Brody get to play this week?

(A) 5 (B) 6 (C) 9 (D) 10

3 Jayden must practice running for $\frac{1}{6}$ hours each day. He has spent $8\frac{4}{6}$ hours practicing. How many days total has Jayden been running if he can only practice $\frac{1}{6}$ hours each day?

(A) 56 days (B) 52 days (C) 48 days (D) 36 days

4 Add $14\frac{11}{35} + 17\frac{8}{35}$

(A) $37\frac{18}{35}$ (B) $34\frac{11}{35}$ (C) $27\frac{11}{35}$ (D) $31\frac{19}{35}$

FRACTIONS

4.4 Mixed Numbers

5. Subtract $27\frac{9}{18} - 22\frac{3}{18}$

A) $9\frac{3}{18}$ B) $4\frac{1}{18}$ C) $5\frac{6}{18}$ D) $7\frac{9}{18}$

6. Given $t + 12\frac{13}{24} = 19\frac{16}{24}$. Find t

A) $7\frac{3}{24}$ B) $8\frac{4}{24}$ C) $9\frac{8}{24}$ D) $10\frac{1}{24}$

7. Given $36\frac{18}{41} - t = 30\frac{18}{41}$. Find t

A) 9   B) 6   C) 4   D) 8

8. Fill in the missing number

$5\frac{\square}{20} + 9\frac{25}{200} = 14\frac{35}{200}$

A) 2   B) 3   C) 6   D) 1

9. Fill in the missing number

$\square\frac{9}{50} + 4\frac{9}{50} = 15\frac{18}{50}$

A) 9   B) 15   C) 11   D) 19

FRACTIONS

Mixed Numbers   **4.4**

10 Gavin bought $5\frac{8}{2}$ pounds of cabbage. He used $3\frac{5}{2}$ pounds for a recipe. How many pounds of cabbage did Gavin have left?

(A) $3\frac{8}{2}$ (B) $2\frac{3}{2}$ (C) $4\frac{5}{2}$ (D) $1\frac{1}{2}$

11 Allison mixed $7\frac{6}{10}$ cans of red paint with $9\frac{7}{10}$ cans of white paint. How many cans of paint did Allison mix altogether?

(A) $14\frac{15}{10}$ (B) $17\frac{13}{10}$ (C) $15\frac{15}{10}$ (D) $16\frac{13}{10}$

12 Jerry must work for $\frac{1}{6}$ hours each day. He has worked for $31\frac{2}{6}$ hours. How many days has Jerry been working?

(A) 188 days   (B) 189 days   (C) 177 days   (D) 187 days

13 Tom had $\$70\frac{8}{55}$. He earned $\$49\frac{13}{55}$ from selling vegetables and $\$26\frac{10}{55}$ from selling fruit.
How much money did Tom have altogether?

(A) $\$121\frac{17}{55}$ (B) $\$119\frac{19}{55}$ (C) $\$145\frac{31}{55}$ (D) $\$96\frac{81}{55}$

FRACTIONS

4.4 Mixed Numbers

14 Given $10\frac{2}{70} + x = 15\frac{26}{70}$. Find x

A) $7\frac{26}{70}$ B) $5\frac{24}{70}$ C) $9\frac{26}{70}$ D) $11\frac{28}{70}$

15 Ian must work for $\frac{1}{4}$ hours each day. He has worked $20\frac{7}{4}$ hours. How many days has Ian been working?

A) 69 days   B) 89 days   C) 77 days   D) 87 days

NEXT CHAPTER:
4.5 Chapter Review

FRACTIONS

4.5 Chapter Review

1. Which fraction can be simplified into another equivalent fraction?

   (A) $\frac{7}{49}$ (B) $\frac{5}{33}$ (C) $\frac{8}{45}$ (D) $\frac{2}{41}$

2. Alice is prepared a apple juice. She pours Olivia a glass with $\frac{5}{12}$ of apple juice and another glass with $\frac{4}{12}$ of the apple juice for herself. Which expression represents the amount of apple juice remaining?

   (A) $\frac{1}{12} + \frac{1}{12}$

   (B) $\frac{1}{12} + \frac{1}{12} + \frac{1}{12} + \frac{1}{12}$

   (C) $\frac{1}{12} + \frac{1}{12} + \frac{1}{12} + \frac{1}{12} + \frac{1}{12}$

   (D) $\frac{1}{12}$

3. Axel owns 15 pairs of shoes and 28 watches. One-fifth of his shoes and one-seventh of his watches are black. Which equation can be used to represent the number of black shoes and black watches Axel owns?

   (A) $\frac{1}{5}(15) = 3$ (B) $\frac{1}{5}(15) = 3$ (C) $\frac{1}{5}(25) = 5$ (D) $\frac{1}{6}(30) = 5$

   $\frac{1}{5}(15) = 3$ $\frac{1}{7}(28) = 4$ $\frac{1}{7}(35) = 5$ $\frac{1}{8}(15) = 3$

109

FRACTIONS

4.5 Chapter Review

4. Sam bought a glass with a capacity of $45\frac{9}{90}$ ounces. The capacity of Tom's glass is 15 times this amount. What fraction represents the capacity of Tom's glass?

(A) $635\frac{9}{90}$ (B) $600\frac{105}{90}$ (C) $675\frac{135}{90}$ (D) $615\frac{2}{90}$

5. Alexandra ate $\frac{9}{20}$ of the watermelon.
What would the equivalent fraction be if the denominator were 200?

6. Mason and Nolan are painting a wall. Mason has painted $\frac{60}{100}$ of the wall. Nolan has painted $\frac{2}{10}$ of the wall. Write an equation that represents the fraction of the walls they have painted.

FRACTIONS

Chapter Review 4.5

7 Use the models to complete the equivalent fraction sentence.

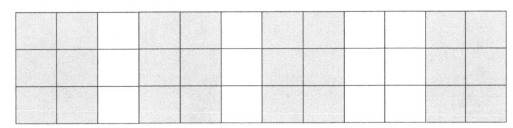

$\frac{8}{12} = \frac{\square}{36}$

Ⓐ $\frac{7}{49}$　　　Ⓑ $\frac{5}{33}$　　　Ⓒ $\frac{8}{45}$　　　Ⓓ $\frac{2}{41}$

8 Give two fractions that are equivalent to $\frac{19}{114}$

_____

_____

FRACTIONS

4.5 Chapter Review

9 Which addition expression represents the total shaded parts in this fraction model?

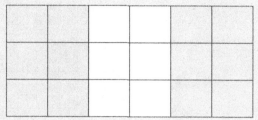

Ⓐ $\dfrac{6}{12} + \dfrac{6}{12}$ Ⓑ $\dfrac{12}{18} + \dfrac{9}{18}$ Ⓒ $\dfrac{6}{18} + \dfrac{9}{18}$ Ⓓ $\dfrac{9}{12} + \dfrac{6}{12}$

10 Use the area model below to answer the question.

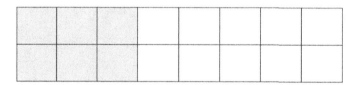

Which expression is shown?

Ⓐ $\dfrac{6}{8} \times \dfrac{1}{2}$ Ⓑ $\dfrac{7}{16} \times \dfrac{1}{16}$ Ⓒ $\dfrac{6}{16} + \dfrac{1}{12}$ Ⓓ $\dfrac{4}{8} + \dfrac{6}{16}$

FRACTIONS

Chapter Review — 4.5

11 $3\frac{2}{3} \times \frac{14}{33} - \frac{1}{9} = \ldots$

(A) $6\frac{2}{33}$ (B) $1\frac{4}{9}$ (C) $2\frac{1}{9}$ (D) $5\frac{3}{33}$

12 $9 \div 4\frac{7}{5} + \frac{2}{3} = \ldots$

(A) $3\frac{4}{5}$ (B) $1\frac{4}{3}$ (C) $2\frac{1}{3}$ (D) $4\frac{3}{5}$

13 Choose the expression that shows adding $7\frac{4}{3} + 8\frac{2}{3}$ using equivalent fractions.

(A) $\frac{4}{5} + \frac{4}{5}$ (B) $\frac{25}{3} + \frac{26}{3}$ (C) $\frac{2}{3} + \frac{2}{3}$ (D) $\frac{3}{35} + \frac{3}{35}$

14 Cruz is running at the rate of $2\frac{2}{5}$ miles per hour. At this rate, how far will Cruz run in $3\frac{8}{4}$ hours?

(A) 3 miles (B) 18 miles (C) 6 miles (D) 12 miles

15 Milo is making a cake that uses $\frac{6}{14}$ cups of flour. If Milo has $15\frac{6}{2}$ cups of flour, how many cakes can he make?

(A) 42 (B) 3 (C) 28 (D) 14

FRACTIONS

4.5 Chapter Review

16. If $5\dfrac{8}{4} \div \dfrac{7}{12} = 4 \div b$. What is the value of b?

(A) 1   (B) 5   (C) 3   (D) 7

17. Lucas needs $3\dfrac{7}{10}$ cups of wheat flour for one whole wheat loaf of bread. How much wheat flour would Lucas need for six whole wheat bread loaves?

(A) $23\dfrac{4}{5}$ (B) $22\dfrac{1}{5}$ (C) $21\dfrac{1}{10}$ (D) $24\dfrac{3}{10}$

18. Calculate $8 - \dfrac{7}{19 - \dfrac{8}{12}} \cdot \dfrac{1}{6} - \dfrac{1}{8}$

(A) $8\dfrac{4}{5}$ (B) $3\dfrac{3}{5}$ (C) $7\dfrac{1}{4}$ (D) $5\dfrac{2}{3}$

19. $7 \div 4\dfrac{5}{4} + 13 =$

(A) $1\dfrac{2}{5}$ (B) $2\dfrac{4}{5}$ (C) $2\dfrac{1}{10}$ (D) $4\dfrac{3}{4}$

FRACTIONS

Chapter Review 4.5

20 The area of the triangle can be calculated using the formula $A = \frac{1}{2} \times$ Base $\times$ Height. In the triangle the base is $3\frac{6}{3}$ cm long. If the area of the triangle is 30cm², What is the height of the triangle?

CHAPTER 5

CONVERSIONS

CONVERSIONS

5.1 Time Across the Hours

Time Across the Hours

Time is a measure that enables us to order sequences of events, compare the duration of events, and determine the intervals between them.

Historically, time was measured using devices such as sundials, where a shadow was cast on a set of markings that were standardized to the hour; water clocks, which could be used to measure the hours after sunset, but required manual timekeeping to replenish the flow of water, and the hourglass, which uses the flow of sand to measure the time. Today, time is measured using clocks that give the time analogically or digitally. The table below shows the conversions for the different units of time. To change from a larger unit to a smaller one, multiply, and to change from a smaller unit to a larger one, divide.

Units of Time	
1 minute (min)	60 second(s) (It takes you around 1 s to say "one thousand one")
1 hour (h)	60 minutes
1 day (d)	24 hours

Example: How many minutes are there in two hours?
Solution: 1 h = 60 min

Adding and subtracting time intervals

CONVERSIONS

Time Across the Hours 5.1

1. The movie began at 10 a.m. and finished at 1:30 p.m. How long was the movie?

   A) 3 hours 30 minutes
   B) 5 hours 30 minutes
   C) 1 hour 30 minutes
   D) 2 hours 30 minutes

2. Convert 4 hours into minutes:

   A) 160 minutes
   B) 240 minutes
   C) 180 minutes
   D) 280 minutes

3. 2 hours 30 minutes + 5 hours 40 minutes

   A) 8 hours 30 minutes
   B) 9 hours 60 minutes
   C) 7 hour 28 minutes
   D) 8 hours 10 minutes

4.  Convert 720 seconds into minutes:

   A) 9 minutes
   B) 24 minutes
   C) 12 minutes
   D) 18 minutes

5. Kelly practiced dancing for 3 hours, while her sister Lily practiced dancing for 132 minutes. Which of the two girls spent more time dancing?

   A) Lily
   B) Kelly

CONVERSIONS

5.1 Time Across the Hours

6 How much time has passed between 7:30 p.m and 11:55 p.m ?

A) 3:55 B) 5:30 C) 4:25 D) 6:55

7 Lexi reached the theater at 12:15 a.m. The movie starts at 1:00 a.m. How many minutes are there until the movie starts?

A) 45 minutes B) 40 minutes
C) 35 minutes D) 25 minutes

8 Randy put a pizza in the oven at 2:05 p.m. At what time should Randy take the pizza out of the oven if the pizza needs 45 minutes to be cooked?

A) 4:55 p.m. B) 3:45 p.m.
C) 2:25 p.m. D) 2:50 p.m.

9 Kevin started to solve the math problems at 1:00 p.m. He finished solving the problems at 3:30 p.m. How long did Kevin take to solve the math problems?

A) 3 hours 40 minutes B) 2 hours 30 minutes
C) 1 hour 20 minutes D) 4 hours 10 minutes

10 Luke arrived at 9:30 a.m at a museum. He left the museum 60 minutes later. What time was it when Luke left the museum?

A) 11:55 a.m. B) 9:50 a.m.
C) 10:30 a.m. D) 10:50 a.m.

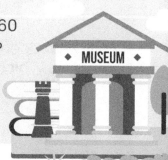

CONVERSIONS

Time Across the Hours 5.1

11 Clark played basketball. He started to play at 2:15 p.m. It was 4:47 p.m. when he finished. How long did Clark play basketball?

A) 2 hours 32 minutes B) 2 hours 37 minutes
C) 3 hour 27 minutes D) 4 hours 15 minutes

12 Demi watched the drama. The drama started at 8:30 p.m. and ended 60 minutes later. What time was it when the drama ended?

A) 8:55 B) 8:30 C) 9:30 D) 7:55

13  Greta arrived at the aquarium at 11:25 a.m. She spent 2 hours and 10 minutes at the aquarium. At what time did Greta leave the aquarium?

A) 2:55 a.m. B) 1:35 p.m.
C) 1:35 a.m. D) 2:20 p.m.

14 Adam spends 40 minutes running every day. How many hours does he spend each year running if he runs every day of the year?

A) 212.6 hours B) 262.8 hours
C) 243.3 hours D) 251.6 hours

15 A TV show was on from 9:00 p.m. to 10:00 p.m. Four times the show was interrupted by advertising commercials, each time for five minutes. How long did the TV show last?

A) 25 minutes B) 45 minutes
C) 30 minutes D) 40 minutes

CONVERSIONS

5.2 Relating Conversions to Place Value

Relating Conversions to Place Value

Prefixes

The most commonly used prefixes in the metric system are shown in the following table:

Prefix Name	Meaning	Symbol
Kilo	1000	k
Centi	$\frac{1}{100}$	c
Milli	$\frac{1}{1000}$	m

Units Conversion

The basic metric unit for measuring distance is the meter. The most commonly used units of length are given in the tables below:

Metric Unit	Conversion
Millimeter (mm)	1 mm = $\frac{1}{10}$ cm = $\frac{1}{100}$ dm = $\frac{1}{1000}$ m
Centimeter (cm)	1 cm = 10 mm = $\frac{1}{10}$ dm = $\frac{1}{100}$ m
Decimeter	1 dm = 100 mm = 10 cm = $\frac{1}{10}$ m
Meter (m)	1 m = 10 dm = 100 cm = 1000 mm
Kilometer (km)	1 km = 1000 m

CONVERSIONS

Relating Conversions to Place Value

Units of weight

Metric units are used to measure the mass of an object. Mass is the amount of matter that an object contains. Most commonly used units of measurement for mass are the milligram (mg), the gram (g), the kilogram (kg), and the metric ton (T).

Units of Weight	Conversion
Milligrams (mg)	$1 \text{ mg} = \frac{1}{1000} \text{ g} = \frac{1}{1000000} \text{ kg}$
Grams (g)	$1 \text{ g} = 1000 \text{ mg} = \frac{1}{1000} \text{ kg}$
Kilogram (kg)	$1 \text{ kg} = 1000 \text{ g} = 1000000 \text{ g}$
Ton (T)	$1 \text{ T} = 1000 \text{ kg}$

Capacity

The capacity of a fluid is measured. The most commonly used metric units for capacity are liters (L), centiliters (cL), and milliliters (mL).
Just like linear units.

Capacity	Conversion
Milliliters (ml)	$1 \text{ mL} = \frac{1}{10} \text{ cL} = \frac{1}{1000} \text{ L}$
Centiliter (cL)	$1 \text{ cL} = 10 \text{ mL} = \frac{1}{1000} \text{ L}$
Liter (L)	$1 \text{ L} = 100 \text{ cL} = 1000 \text{ mL}$

CONVERSIONS

5.2 Relating Conversions to Place Value

1. Each day, Tara drives 7 kilometers and 5,000 meters. In all, how many meters does Tara drive each day?

   (A) 10,000 m   (B) 12,000 m   (C) 11,000 m   (D) 7,000 m

2. Amos measured a long ribbon. It is 500 millimeters long. How long is the ribbon in centimeters?

   (A) 50 cm   (B) 100 cm   (C) 40 cm   (D) 80 cm

3. Judith has five pieces of wire that are 85 cm long each. How much wire does she have in total, in millimeters?

   (A) 4,850 mm   (B) 4,855 mm   (C) 5,375 mm   (D) 4,250 mm

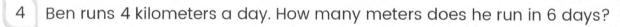

4. Ben runs 4 kilometers a day. How many meters does he run in 6 days?

   (A) 28,000 m   (B) 26,000 m   (C) 24,000 m   (D) 20,000 m

5.  Axel's study table is 105 centimeters long and 55 centimeters wide. How many more millimeters is the length of the study table than its width?

   (A) 600 mm   (B) 500 mm   (C) 400 mm   (D) 300 mm

CONVERSIONS

Relating Conversions to Place Value — 5.2

6. A bag contains 8 packs of salt. The total mass of all eight packs is 16 kg. What is the mass of each pack in grams?

A) 2,000 g   B) 4,000 g   C) 6,000 g   D) 8,000 g

7. If one pencil has a mass of 2 g and 500 pencils have a mass of 1 kg how many kilograms are 6,000 pencils?

A) 6 kg   B) 12 kg   C) 18 kg   D) 24 kg

8. One carrot weighs 13 g. What is the weight of 15 boxes of 20 carrots, in kilograms and grams?

A) 4 kg 400 g   B) 1 kg 600 g   C) 3 kg 900 g   D) 2 kg 700 g

9.  Each cupcake in a box weighs 75 g, and the entire box weighs 3 kg. How many cupcakes are in the box?

A) 60   B) 40   C) 50   D) 30

10. A barrel contains 180 liters of water. During the day, 110 liters and 200 ml of water are used. How much water is left in the barrel?

A) 34 L 200 ml   B) 46 L 500 ml
C) 57 L 700 ml   D) 69 L 800 ml

CONVERSIONS

5.2 Relating Conversions to Place Value

11. One stone weighs 35 kilograms and 250 grams. Another stone weighs 38 kilograms and 120 grams. What is the total weight, in milligrams, of the two stones?

(A) 69,170,000 mg (B) 73,470,000 mg
(C) 73,370,000 mg (D) 81,370,000 mg

12. Write 11 L 710 ml in millilitres:

(A) 11,710 ml (B) 110,710 ml (C) 11,7010 ml (D) 11,700 ml

13. Vale prepared 4 liters of apple juice. She wanted each of her 10 friends to have 2 glasses of 150 mL of apple juice. Did she prepare enough? Explain.

(A) She did not prepare enough (B) She prepared enough

14. Dylan weighs 42 kilograms and 800 grams. Felipe weighs 4980 g less than Dylan. Quentin weighs 2400 g less than Felipe. How much does Quentin weigh?

(A) 21,870 g (B) 29,350 g (C) 35,420 g (D) 39,210 g

15. Rosie has three curtains. The first curtain is 5 m 20 cm long. The second curtain is 56 cm shorter than the first. The third curtain is 157 cm longer than the second curtain. What is the difference in millimeters between the length of the first curtain and the third curtain?

(A) 1,090 mm (B) 1,037 mm (C) 1,056 mm (D) 1,010 mm

CONVERSIONS

5.3 Centimetres and Meters

Centimetres and Meters

There are different systems for measuring distances. A commonly used system is the metric system. The units often used are the kilometer (km), the meter (m), the centimeter (cm), and the millimeter (mm).

- The distance between cities is measured in kilometers.
- The height of a building is measured in meters.
- The length of your arm is measured in centimeters.
- The thickness of 10 sheets of paper is about 1 millimeter.

1 km = 1,000 m
1 m = 100 cm
1 cm = 10 mm
1 m = 1,000 mm

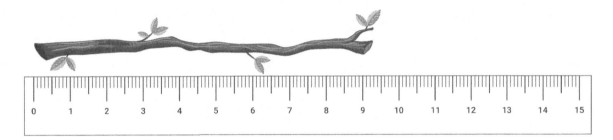

The length of the stick under the ruler is between 9 cm and 10 cm.
To the nearest cm, it is 9 cm.
The exact length of the stick is 9 cm and 2 mm or simply 92 mm.

CONVERSIONS

Centimetres and Meters

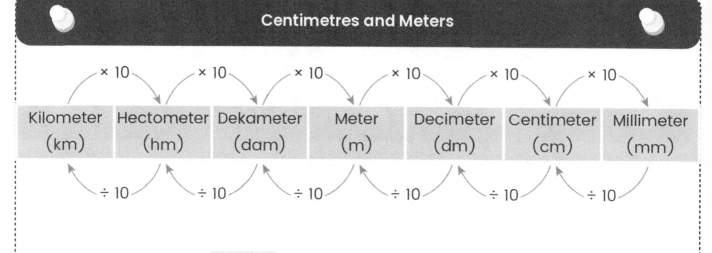

30 mm or 3 cm

The length of the paperclip measures 3 cm or 30 mm.

3 cm is equivalent to 3×10 =30 mm.

CONVERSIONS

Centimetres and Meters — 5.3

1. Abilene planted a fruit tree in her garden when it was 24 centimeters and 8 millimeters tall. Since that time, the tree has grown 8 centimeters and 1 millimeter.
What is the height of the tree now?

2. One day Bobby painted 2 meters and 45 centimeters of a long gate. On the next day, he painted 5 meters and 20 centimeters of this gate.
What is the total length of the gate that Bobby painted?

3. Jasmine's room window is 4 meters tall. Lily's room window is 12 centimeters shorter than Jasmine's room window. How tall is the window in Lily's room?

4. The length of a photo frame is 34 centimeters. What is the length of the photo frame in millimeters?

   A) 380 mm   B) 310 mm   C) 340 mm   D) 350 mm

CONVERSIONS

5.3 Centimetres and Meters

5. The height of an iron rod is 7 meters. What is the height of this rod in millimeters?

- A) 4,500 mm
- B) 5,000 mm
- C) 6,500 mm
- D) 7,000 mm

6. Christian ran 18 km 30 m in 2 hours. What was the distance, in meters, Christian ran per hour if he ran the same distance each hour?

- A) 9,005 mm
- B) 9,015 mm
- C) 6,1 mm
- D) 9,540 mm

7. Stacy is 2 meters and 54 centimeters tall. Express Stacy's height in millimeters.

- A) 1,990 mm
- B) 2,011 mm
- C) 2,540 mm
- D) 2,970 mm

8. The length of a rabbit is 92 millimeters. A second rabbit is 5 centimeters longer. What is the length of the second rabbit, in millimeters?

- A) 1,990 mm
- B) 2,011 mm
- C) 2,540 mm
- D) 2,970 mm

9. The length of a pillar is 6 m and 64 cm. What is the length of the pillar in millimeters?

- A) 6,240 mm
- B) 6,640 mm
- C) 7,240 mm
- D) 7,950 mm

10. Aldo requires a steel rod with a diameter of 30 cm. Aldo is using a catalog that lists sizes in millimeters. What size must Aldo order?

- A) 250 mm
- B) 200 mm
- C) 350 mm
- D) 300 mm

CONVERSIONS

Centimetres and Meters 5.3

11 A bird is flying at an elevation of 12 km. How many meters high is the bird flying?

(A) 12,000 m (B) 6,000 m (C) 18,000 m (D) 10,000 m

12 The width of the living room is 5 meters. What is the width of the living room in centimeters?

(A) 300 cm (B) 400 cm (C) 500 cm (D) 600 cm

13  The apple tree is 580 cm tall, and the orange tree is 5,160 mm tall. Which is taller, the apple tree or the orange tree?

(A) Apple tree is taller (B) Orange tree is taller

14 Elliot walked for 4 km, and Hector walked for 1800 m. Who walked more, Elliot or Hector?

(A) Hector walked more (B) Elliot walked more

15 Which is greater 8 km or 860,000 mm ?

NEXT CHAPTER:
5.4 Grams and Kilograms

CONVERSIONS

5.4 Grams and Kilograms

Grams and Kilograms

Units of weight

Metric units are used to measure the mass of an object. Mass is the amount of matter that an object contains. The most commonly used units of measurement for mass are the milligram (mg), the gram (g), the kilogram (kg), and the metric ton (T). The mass of an object can be measured with scales or a pan balance.

The mass of a cargo ship is measured in metric tons.

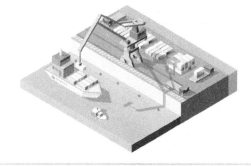

The mass of a human is measured in kilograms.

The mass of a bar of candy is measured in grams.

The mass of a pinch of salt is measured in milligrams.

CONVERSIONS

Grams and Kilograms

The table below shows how to convert mass from one unit to another. As with units of length, when converting from a larger unit to a smaller unit, we multiply by powers of 10. When converting from a smaller unit to a larger unit, we divide by powers of 10.

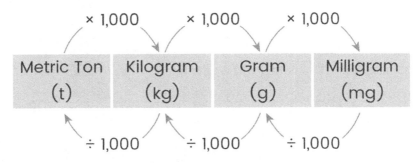

Example:
How many grams are there in 4 kg?
Solution:
4 × 1,000 = 4,000 kg

---

1. Each toothbrush weighs 6 g. There are 12 toothbrushes in a pack. What is the total weight of the toothbrush in 40 packs in kilograms?

   (A) 2 kg 420 g   (B) 2 kg 880 g   (C) 3 kg 930 g   (D) 4 kg 780 g

2. A black bag weighs 450 grams. A white bag weighs 900 grams more than a black bag. What is the weight of the white bag?

   (A) 1 kg   (B) 2 kg 110 g   (C) 2 kg   (D) 1 kg 350 g

CONVERSIONS

5.4 Grams and Kilograms

3 Mary bought a red onion and some corn from the market. The mass of the red onion was 625 grams, and the mass of the corn was 2 kg and 400 grams. How much greater was the mass of the corn than the mass of the red onion?

A) 1,090 g   B) 1,237 g   C) 1,775 g   D) 1,984 g

4 The weight of a mouse is 2 kilograms. What is the weight of the mouse in grams?

A) 2,000 g   B) 2,500 g   C) 3,000 g   D) 3,500 g

5 A mirror weighs 14 kg and 275 grams. What is the weight of the mirror in grams?

A) 28,275 g   B) 14,275 g   C) 10,270 g   D) 7,250 g

6 A piece of cake weighs 60 grams. How many milligrams are equal to 60 grams?

A) 60,000 mg   B) 6,000 mg   C) 600 mg   D) 600,000 mg

7 Carl has 80 pears, each with a mass of 30 grams. What is the total mass, in kilograms, of Carl's pears?

A) 1 kg   B) 2 kg   C) 2 kg 400 g   D) 1 kg 850 g

8 Which is lighter, a pumpkin that weighs 3,800 grams or three cauliflowers with a mass of 1 kilogram each?

A) Cauliflower is lighter   B) Pumpkin is lighter

CONVERSIONS

Grams and Kilograms 5.4

9 A dog weighs 13 kilograms, and a cat weighs 8,400 grams. Which is heavier, the dog or the cat?

- (A) A cat is heavier
- (B) A dog is heavier

10 A fruit shop has 62 kg and 250 grams of plums. 34 kilograms and 170 grams are sold. How much is left?

- (A) 28 kg 80 g
- (B) 24 kg
- (C) 29 kg 400 g
- (D) 21 kg

11 One eraser weighs 2 grams. What do 250 erasers weigh?

- (A) 150 g
- (B) 600 g
- (C) 250 g
- (D) 500 g

12 Eight pineapples weigh 10 kg 640 g. What does one pineapple weigh?

- (A) 3 kg
- (B) 1 kg 330 g
- (C) 2 kg 400 g
- (D) 2 kg

13 Gil buys two baskets of apricots. There is a large basket holding 32 kg and a small basket holding 16 kg and 200 g. What is the total weight of the apricots he has bought?

- (A) 46 kg 100 g
- (B) 46 kg 330 g
- (C) 48 kg 200 g
- (D) 48 kg 600 g

14 Five watermelons weigh 12 kg 120 g. What does one watermelon weigh?

- (A) 2 kg 424g
- (B) 2 kg 330 g
- (C) 2 kg 400 g
- (D) 2 kg 200g

CONVERSIONS

5.4 Grams and Kilograms

15. Nick plucked 7 kg and 458 g of mangoes. Royce plucked 674 grams more. How many mangoes did Royce pluck?

(A) 6 kg 100 g (B) 4 kg 326 g (C) 8 kg 132 g (D) 6 kg 428 g

NEXT CHAPTER:
5.5 Liters and Milliliters

CONVERSIONS

5.5 Liters and Milliliters

Liters and Milliliters

The volume of a solid is the amount of space it occupies. Its capacity is the amount of fluid it can hold. The most commonly used metric units for capacity are liter (L) and milliliter (mL). You can measure capacity using a measuring cylinder or jug, although there are other devices.

The amount of water in a bathtub is measured in liters.

The amount of juice in a glass is measured in milliliters.

The amount of medicine in a spoon is measured in milliliters.

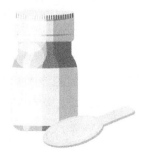

CONVERSIONS

Liters and Milliliters

People often make incorrect distinctions when referring to the amount of matter that a container holds.

We can say "the volume of this glass is 200 mL", but it is more accurate to say "the capacity of the glass is 200 mL".

There is a close relation between units of capacity and units of volume.

Relation between Volume and Capacity
1 L = 1,000 ml
1,000 L = 1 m^3
1 L = 1,000 cm^3
1 mL = 1 cm^3

Note that 1,000 L is indeed 1,000,000 mL

Example:

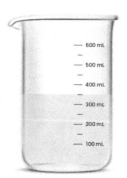

The volume of an object can be measured by finding the amount of liquid that the object displaces when it is totally submerged. The water level shows 350 mL before the orange was submerged. After immersing the orange, the water level rose to 550 mL. The amount of water displaced by the orange is   mL. Therefore, the volume of this orange is 200 mL.

CONVERSIONS

Liters and Milliliters

Therefore, the volume of this orange is 200 mL. In fact, the volume of the orange would be better represented as 200 cm³.
If a cube of side 1 cm is totally submerged in the medicine cup shown to the right, the water level must increase by 1 mL.

1. A paint can has a capacity of 5 liters. Mark is pouring the contents of the can into a 5000 ml paint can. Will the paint can overflow?

_____

_____

2. Drew has two small water tanks. The volume of one is 48 liters, and the volume of the other is 72 liters. How much water is needed to fill both water tanks? Give an answer in milliliters.

_____

_____

CONVERSIONS

5.5 Liters and Milliliters

3. A barrel currently contains 46,800 milliliters of water. An additional 34,200 milliliters of water are poured into the barrel. How many liters of water are there in the barrel now?

4. The capacity of a fish tank is 10 liters. Express this capacity in milliliters.

   (A) 100,000 ml (B) 10,000 ml (C) 1,000 ml (D) 100 ml

   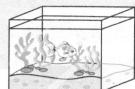

5. An empty bucket with a capacity of 18 L is to be filled with a number of glasses of water each with capacity 250 ml. How many glasses are required to fill the empty bucket with water?

   (A) 60 (B) 64 (C) 70 (D) 72

6. A paint can hold 4 L and 500 ml. 3 L and 300 ml are used. 450 ml are added. How much paint is in the paint can?

   (A) 1 L 650 ml (B) 2 L 150 ml (C) 1 L 980 ml (D) 2 L 650 ml

7. Kiara buys a 3 L bottle of water. Three-ninths are used. A quarter of what is left is used. How much water is left?

   (A) 1,300 ml (B) 1,400 ml (C) 1,500 ml (D) 1,600 ml

CONVERSIONS

Liters and Milliliters   **5.5**

8. Gemma makes 2 liters and 700 ml of carrot juice. She pours one-third into a jug, and the rest is shared equally among five glasses. How much juice is in each glass?

9. On Friday, Derrick prepared 24 liters of orange juice in his shop. On Saturday, he prepared 1480 ml more than he had the day before. How much orange juice does he prepare on Saturday?

10. One coffee cup holds 250 ml. What do nine cups hold?

    (A) 3 L 650 ml   (B) 2 L 250 ml   (C) 4 L 750 ml   (D) 3 L 650 ml

11. One water bottle holds 1,050 ml. How much water is in twenty-one bottles?

    (A) 21 L 150 ml   (B) 20 L 50 ml   (C) 21 L 50 ml   (D) 22 L 50 ml

12. A bottle of soybean oil holds 260 ml. How many bottles can be filled with 5 L and 980 ml?

    (A) 21   (B) 27   (C) 23   (D) 26

CONVERSIONS

5.5 Liters and Milliliters

13 Wine is made by adding 1720 ml of grape juice to 3 liters of water. The drink is equally poured into sixteen glasses. How much wine is in each glass?

(A) 3 L 650 ml (B) 2 L 250 ml (C) 4 L 750 ml (D) 3 L 650 ml

14 Jami prepares 10,000 millilitres of orange juice to sell. She bottles the orange juice in small bottles that hold a $\frac{1}{5}$ litre each. How many bottles does Jami need?

(A) 40 (B) 70 (C) 50 (D) 60

15 At a wedding, the 92 guests are each served a 320 ml glass of water. Approximately, how many 2 L bottles are needed?

(A) 20 (B) 15 (C) 18 (D) 21

NEXT CHAPTER:
5.6 Chapter Review

CONVERSIONS

5.6 Chapter Review

1. Diego's soccer practice began at 1:30 p.m. The team practiced offense for 1 hour and 20 minutes and defense for 1 hour and 50 minutes. What time did Diego's soccer practice end?

A) 4:30 p.m.   B) 3:50 p.m.   C) 2:40 p.m.   D) 3:20 p.m.

2. Joseph works at a burger shop. Each morning, Joseph walks 15 minutes to the bus stop. The bus ride to the shop takes 45 minutes. If Joseph works six days per week, how many hours does he spend traveling to his job at the burger shop?

A) 5 hours 45 minutes   B) 6 hours
C) 7 hours   D) 8 hours 15 minutes

3. Levi is making a pizza and needs 3 kg of the flour. He looks in his pantry and discovers that he only has 1 kg 95 g. When he goes to the store, he discovers that they measure flour in grams. How many grams of flour should he order?

A) 1,890 g   B) 1,950 g   C) 1,905 g   D) 1,814 g

4. Maria had six identical bottles, each containing 450 milliliters of milk. She poured the milk from them evenly into five glasses. What is the volume of milk in each of the glasses?

A) 300 ml   B) 480 ml   C) 350 ml   D) 540 ml

CONVERSIONS

5.6 Chapter Review

5. Kobe has a pot of water. He does not know how much water he has, but he knows he can fill 10 glasses that have a capacity of 160 millimeters each. How much water does he have?

   Kobe has _____ of water.

6. It is 3:25 P.M. What will the time be after 1 hours and 50 minutes?

   (A) 5:20 p.m.   (B) 5:15 p.m.   (C) 5:40 p.m.   (D) 5:25 p.m.

7. What time is 3 hours and a half before 6:50 in the evening?

   (A) 4:20 p.m.   (B) 2:15 p.m.   (C) 3:20 p.m.   (D) 3:40 p.m.

8. Which list shows the measurements in order from greatest to least?

   (A) 4 days, 30 hours, 1260 minutes, 700 seconds
   (B) 7 hours, 9 days, 45 minutes, 240 seconds
   (C) 1 day, 35 hours, 124 minutes, 5490 seconds
   (D) 3450 seconds, 12 days, 450 minutes, 34 hours

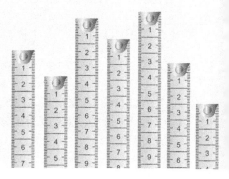

9. Convert 3840 seconds into minutes

   (A) 11 minutes   (B) 42 minutes
   (C) 19 minutes   (D) 64 minutes

CONVERSIONS

Chapter Review 5.6

10. James played Ice hockey. He started to play at 2:35 p.m. It was 5:50 p.m. when he finished. How long did he play?

   A) 1 hours 35 minutes
   B) 3 hours 15 minutes
   C) 4 hour 25 minutes
   D) 2 hours 45 minutes

11. A box contains 12 ice cream bars. The total mass of all 12 ice cream bars is 48 kg. What is the mass of each ice cream bar in grams?

   A) 4,000 g
   B) 8,000 g
   C) 12,000 g
   D) 16,000 g

12. One table weighs 12 kilograms and 150 grams. Another table weighs 18 kilograms 160 grams. What is the total weight, in milligrams, of the two tables?

   A) 47,610,000 mg
   B) 10,820,000 mg
   C) 30,310,000 mg
   D) 25,680,000 mg

13. Mayson has three ladders. The first ladder is 6 m 80 cm long. The second ladder has a length 70 cm less than the first. The third ladder is 184 cm longer than the second ladder. What is the difference in millimeters between the length of the second ladder and the third ladder?

   A) 1840 mm
   B) 1767 mm
   C) 1098 mm
   D) 1821 mm

14. The length of the straw is 58 centimeters. What is the length of the straw in millimeters?

   A) 460 mm
   B) 370 mm
   C) 580 mm
   D) 300 mm

CONVERSIONS

5.6 Chapter Review

15. The width of the wall is 11 meters. What is the width of the wall in centimeters?
    - A) 1,001 cm
    - B) 1,100 cm
    - C) 1,011 cm
    - D) 1,111 cm

16. An iron rod is 290 cm tall and a water pipe is 1,240 mm tall. Which is taller, the iron rod or a water pipe?
    - A) Water pipe is taller
    - B) Iron rod is taller

17. The weight of a baby horse is 45 kilograms. What is the weight of the baby horse in grams?
    - A) 45,000 g
    - B) 54,500 g
    - C) 4,000 g
    - D) 5,500 g

18. Ten red cabbages weigh 16 kg 140 g. What does one red cabbage weigh?
    - A) 2 kg
    - B) 1 kg 140 g
    - C) 1 kg 614 g
    - D) 1 kg

19. An oil can hold 5 liters and 100 ml. 2 liters and 700 ml are used. 670 ml are added. How much oil is in the oil can?
    - A) 3 L 70 ml
    - B) 2 L 50 ml
    - C) 1 L 80 ml
    - D) 4 L 50 ml

20. A juice bottle holds 350 ml. How many bottles can be filled from 4 L and 550 ml?
    - A) 10
    - B) 13
    - C) 19
    - D) 23

CHAPTER 6
GEOMETRIC MEASUREMENT

GEOMETRIC MEASUREMENT

6.1 Area of Rectilinear

Area of Rectilinear

The area of a closed planar figure is the space it covers. The area is measured in units called square units. If the dimensions of the figure are given in meters, then its area is measured in square meters (m^2).

In general, whatever the units used for the dimensions of the figure, its area is measured in square units.

The area of a square of side s is s×s. If the side of a square is one meter then its area would be 1×1 = $1m^2$

In the below figure, if each small square has an area of 1 cm^2 then the area of one row would be 7 cm^2.

 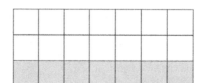 

GEOMETRIC MEASUREMENT

Area of Rectilinear

The area of the whole rectangle in this case is: $3 \times 7 = 21$ cm^2 as there are 3 rows of small squares in the rectangle. But if each small square has an area of 1 cm, then the side of each small square must be 1 cm in length. And the length of the whole rectangle is 7 cm while its width is 3 cm. Its area can then be obtained as the product of its length and its width: $3 \times 7 = 21$ cm^2.

The area of a rectangle = length × width.

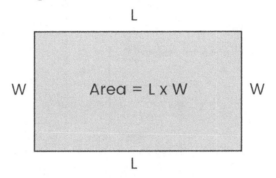

The appropriate unit for the area of a house is m^2.
The appropriate unit for the land area of a country is km^2.
The appropriate unit for the area of a piece of land is a hectare (ha).

Example:
Find the area of the A- Shaped figure without having to count every single square

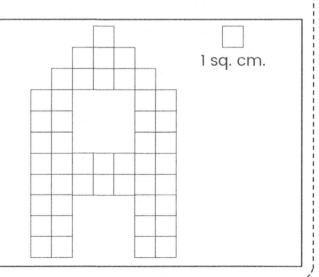

GEOMETRIC MEASUREMENT

Area of Rectilinear

The shape is divided into four recyangles and three small squares leftover. Their areas are:
2 × 8 = 16 square centimeters
2 × 3 = 6 square centimeters
2 × 3 = 6 square centimeters
2 × 8 = 16 square centimeters
There are 3 additional squares of 1 square centimeter each.
The area of the shape is
16+6+6+16+3 = 22+6+16+3
= 47 square centimeters

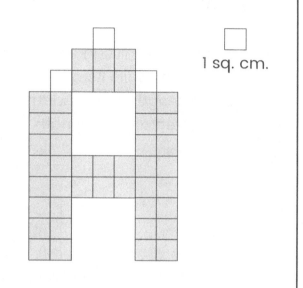

1 sq. cm.

1. What is the area in units of the unshared shape when the larger square has a side length of 10 units and the small square a side length of 2 units?

   A) 96 square units B) 94 square units
   C) 88 square units D) 90 square units

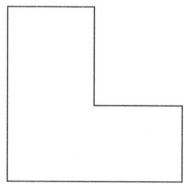

GEOMETRIC MEASUREMENT

Area of Rectilinear   **6.1**

2   Find the area of the figure given that each unit on the grid is 1 cm in length.

A) 28 cm²
B) 22 cm²
C) 26 cm²
D) 24 cm²

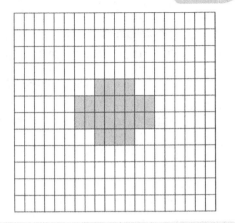

3   The area of this square ☐ is 18 square units. What is the area of this shape?

A) 124 un²
B) 100 un²
C) 144 un²
D) 168 un²

4   Explain how you can find the area of the whole rectangle.

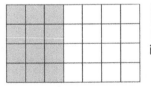

1 sq. inch.

_____

5   Find the area of the given figure:

12 m
4 m
10 m
2 m

A) 74 m²
B) 68 m²
C) 44 m²
D) 54 m²

GEOMETRIC MEASUREMENT

6.1 Area of Rectilinear

6 Cleo wants to cover the floor shown below with tiles. Each tile is one square meter.

Which number sentences can be used to calculate the number of tiles required by Cleo to cover the entire floor? Select the two correct answers.

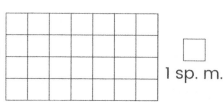
1 sp. m.

- (A) 4×7
- (B) 7+7
- (C) 7×4
- (D) 4+7+7

7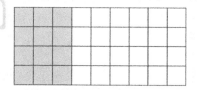

1 sq. inch.

Write a multiplication sentence that gives the area of the unshaded part of the rectangle.

_____

8 What is the area in units of the unshared shape?

9 Find the area of the figure given that each unit on the grid is 1 cm in length.

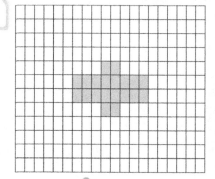

- (A) 28 cm²
- (B) 22 cm²
- (C) 26 cm²
- (D) 24 cm²

GEOMETRIC MEASUREMENT

Area of Rectilinear   6.1

10 Find the area of the figure given that each unit on the grid is 1 cm in length.

     (A) 28 cm²   (B) 22 cm²
     (C) 26 cm²   (D) 24 cm²

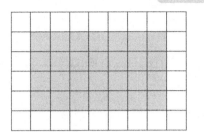

11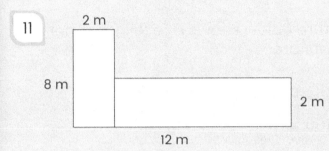

Find the area of the given figure:

     (A) 34 m²   (B) 18 m²
     (C) 44 m²   (D) 36 m²

12 A rectangle is 15 ft long and 4 ft wide. Find the area of the new rectangle whose dimensions are triple in size.

     (A) 448 cm²   (B) 540 cm²
     (C) 260 cm²   (D) 524 cm²

13 James has a garden shaped like an L. What is the area of the garden?

     (A) 70 m²   (B) 50 m²
     (C) 60 m²   (D) 40 m²

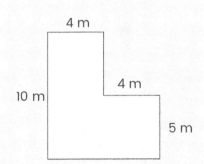

153

GEOMETRIC MEASUREMENT

6.1 Area of Rectilinear

14. The picture below shows the design for parking lot D. What is the area of the lot? Show your work.

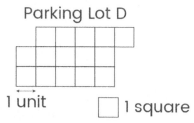

Parking Lot D

1 unit

☐ 1 square

15.

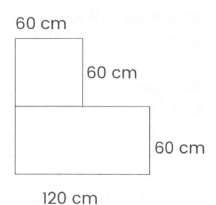

60 cm
60 cm
60 cm
120 cm

The picture below shows a rectilinear shelf. What is its area?

A) 10800 cm² B) 12600 cm²
C) 20800 cm² D) 22600 cm²

NEXT CHAPTER:
6.2 Fixed Area - Varying Perimeter

GEOMETRIC MEASUREMENT

6.2 Fixed Area - Varying Perimeter

Fixed Area - Varying Perimeter

Width

Length

The perimeter of a rectangle is given by: P = 2 × (Width + Length).
The area of a rectangle is given by: A = Width × Length.

Example:

Look at the figures below.

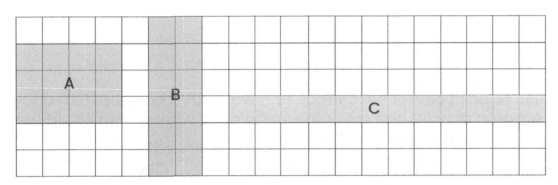

☐ = 1 square meter

GEOMETRIC MEASUREMENT

Fixed Area – Varying Perimeter

a. Compare the areas of shapes A and B.
   The area of A is 3×4 = 12 square meters.
   The area of B is 6×2 = 12 square meters.
   A and B have the same area.

b. Compare the perimeters of A and B.
   Perimeter of A is 4 + 3 + 4 + 3 = 14 meters.
   Perimeter of B is 2 + 6 + 2 + 6 = 16 meters.
   The perimeter of A is less than the perimeter of B.

c. Compare the areas of B and C.
   The area of C is 4×4 = 16 square inches.

d. Compare the perimeters of B and C.
   Perimeter of C is 4 + 4 + 4 + 4 = 16 in.
   B and C have the same perimeter.

---

1. A gallon of paint covers the surface of a wall with a height of 3 m and length 14 m. If a gallon of paint can cover the same area, what length would the wall need to be with a height of 6 m?

(A) 5 m (B) 6 m (C) 9 m (D) 7 m

2. Determine the width and perimeter of a rectangle with a length of 20 cm and an area of 200 cm².

Width = _____

Perimeter = _____

GEOMETRIC MEASUREMENT

Fixed Area – Varying Perimeter   6.2

3 The diagram represents a rectangle 12 cm by 3 cm. How long is the side of a square the same area as this rectangle?

     12 cm
     3 cm

- A) 5 cm
- B) 6 cm
- C) 7 cm
- D) 8 cm

4 Find the length and perimeter of a rectangle for the given area 130 cm² with the width as 10 cm.

Length = _____

Perimeter = _____

5 If you made a square with the same area, what would be the length of each side?

4 cm
4 cm

6 A rectangular walkway has an area of 40 ft². What is the length of the walkway if the width is 5?

- A) 7 ft
- B) 8 ft
- C) 9 ft
- D) 10 ft

GEOMETRIC MEASUREMENT

6.2 Fixed Area - Varying Perimeter

7 Tara wants to plant a rectangular fruit garden with an area of 63 square yards. What is the garden's width if the length is 7 yards?

- (A) 7 yards
- (B) 12 yards
- (C) 8 yards
- (D) 9 yards

8 A rectangular room is 6 m long and has an area of 42 m². What is the perimeter of the room?

- (A) 24 m
- (B) 26 m
- (C) 49 m
- (D) 36 m

9 A playground has a length of 40 m and an area 920 m². What is the perimeter of the playground?

- (A) 224 m
- (B) 226 m
- (C) 136 m
- (D) 126 m

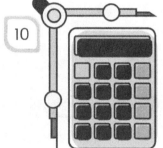

10 A picture frame in a rectangular shape has an area of 810 square cms. It is 90 centimeters long. What is its perimeter?

- (A) 180 cm
- (B) 90 cm
- (C) 198 cm
- (D) 190 cm

11 A rectangular garden has an area of 117 ft². What is the length of the rectangular garden if the width is 13 ft?

- (A) 6 ft
- (B) 10 ft
- (C) 8 ft
- (D) 9 ft

GEOMETRIC MEASUREMENT

Fixed Area - Varying Perimeter 6.2

12 A factory is 50 m long and has an area of 4000 m. Its warehouse has the same area but is 30 m longer. What is the perimeter of the warehouse?

- (A) 260 m
- (B) 230 m
- (C) 220 m
- (D) 240 m

13 Liam's bedroom is 8 m long and 2 m wide. Owen's room has the same area but is 2 m wider. How long is Owen's room?

- (A) 8 m
- (B) 4 m
- (C) 2 m
- (D) 6 m

14 Noah and Logan both calculate rectangles with an area of 72 cm². Noah's rectangle is 12 cm long. Logan's rectangle is 4 cm shorter. Give the length and width of both rectangles.

Noah's rectangle = _____

Logan's rectangle = _____

15 A square has an area of 18 cm². Calculate two different rectangles with the same area as the square. Find the perimeter of each shape. Which shape has the shortest perimeter?

NEXT CHAPTER:
6.3 Fixed Area - Varying Area

GEOMETRIC MEASUREMENT

6.3 Fixed Perimeter - Varying Area

Fixed Perimeter - Varying Area

The perimeter of a rectangle is given by:
$P = 2 \times (\text{Width} + \text{Length})$.

The area of a rectangle is given by:
$A = \text{Width} \times \text{Length}$

Perimeter

The perimeter of a closed planar figure is the length of its boundary or outer edge.

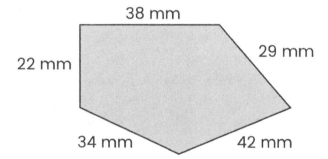

The perimeter of the adjacent figure is:
$38 + 29 + 34 + 42 + 22 = 165$ mm

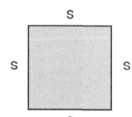

GEOMETRIC MEASUREMENT

Fixed Perimeter - Varying Area

In a square, all sides are equal.
perimeter is s + s + s + s = 4 × s.

For a rectangle, the perimeter is
L + W + W + L = 2 × (W + L).

Example:
Find the perimeter of the rectangle of length 14 cm and width 8 cm.

Solution:
For a rectangle, perimeter = 2 × (Length + Width)
= 2 × (14 + 8) = 2 × (22) = 44 cm.

1. Determine the width and area of a rectangle that has a length 4 cm and perimeter of 20 cm.

   Width = _____

   Perimeter = _____

GEOMETRIC MEASUREMENT

6.3 Fixed Perimeter - Varying Area

2 If you made a square with the same perimeter, what would be the length of each side? Find the area of the square.

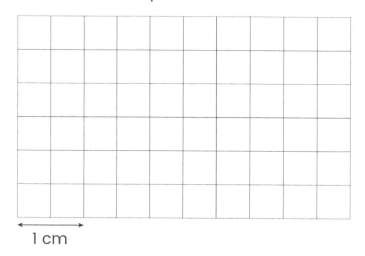

1 cm

Length of each side = _____ cm

Area of the square = _____ cm^2

3 A rectangular room is 6 long and has a perimeter of 30. What is the area of the room?

The area of the room = _____ m^2

4 A rectangular field is three times as long as it is wide. The fence around it is 320 m long. What is the area of the field?

(A) 3200 m^2 (B) 4800 m^2
(C) 2400 m^2 (D) 4000 m^2

GEOMETRIC MEASUREMENT

Fixed Perimeter – Varying Area   **6.3**

5. What is the area of a rectangle 6 feet wide if its perimeter is 60 feet?

   - A) 144 ft²
   - B) 124 ft²
   - C) 134 ft²
   - D) 104 ft²

6. The lengths of the sides of a triangle are 8 cm, 3 cm, and x cm. The perimeter of the triangle is 18 cm. Which of the following equations can be used to find the value of x? Select the two correct answers.

   - A) $8+3=x$
   - B) $8+3+18=x$
   - C) $8+3+x=18$
   - D) $x=18-8-3$

7. The floor of Dona's room is shaped like a rectangle. It is 6 feet wide and 8 feet long. What is the perimeter of the floor of Dona's room?

   - A) 18 feet
   - B) 36 feet
   - C) 28 feet
   - D) 44 feet

8. The perimeter of a square is 99 inches. The length of its side is x inches. Write the equation that can be used to find the value of x?

   _____

9. Cruz and his friends are making rectangular photo frames.
   - The width of each frame is 6 inches.
   - The perimeter of each frame is 22 inches.

   What is the length of each of the frames?

   - A) 5 inches
   - B) 6 inches
   - C) 7 inches
   - D) 8 inches

GEOMETRIC MEASUREMENT

6.3 Fixed Perimeter – Varying Area

10 A rectangular walkway has a perimeter of 180 ft. What is the area of the walkway if the width is 16 ft?

　Ⓐ 1004 ft²　　　　　　　Ⓑ 1184 ft²

　Ⓒ 1224 ft²　　　　　　　Ⓓ 2004 ft²

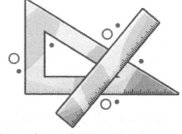

11 The length of a rectangular field is two times longer than its width. The fence around it is 240. What is the area of the field?

12 The basketball court is shown below.

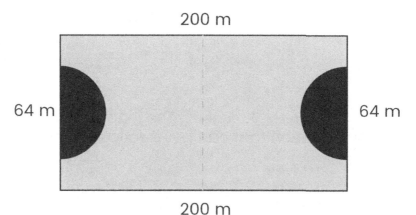

The basketball court has the same perimeter but is 20 m shorter. What is the area of the basketball court?

GEOMETRIC MEASUREMENT

Fixed Perimeter - Varying Area 6.3

13 The perimeter of the square photo is 40 inches. Angie decided to enlarge the photo by doubling the sides. What will the new area be?

- (A) 400 square inches
- (B) 200 square inches
- (C) 300 square inches
- (D) 100 square inches

14 A small rectangle with a width of 5 cm. The perimeter of the small rectangle is 50 cm. What is the length of the square?

- (A) 20 cm
- (B) 40 cm
- (C) 30 cm
- (D) 25 cm

15 A factory is 74 m long and has a perimeter of 248 m. Its warehouse has the same perimeter but is 6 m longer. What is the area of the factory? What is the area of the warehouse?

NEXT CHAPTER:
6.4 Perimeter and Area of Rectangles

GEOMETRIC MEASUREMENT

6.4 Perimeter and Area of Rectangles

Perimeter and Area of Rectangles

The perimeter of a rectangle is given by:
$$P = 2 \times (\text{Width} + \text{Length}).$$
The area of a rectangle is given by:
$$A = \text{Width} \times \text{Length}$$

Example:

Jessica wants to replace the floor tiles in her rectangular room.

a. What information does she need in order to know the number of tiles to buy?

Jessica needs to know the area of the room to know how many tiles to buy.

b. What is the width of the room if the area is 36 m² and the length is 6 m? What can we say about the shape of the room?

The length of the room is 36÷6=6 meters long.

We can say that the room is square.

GEOMETRIC MEASUREMENT

Perimeter and Area of Rectangles 6.4

1. Which number sentences can be used to find the area of the rectangle ABCD? Select the two correct answers.

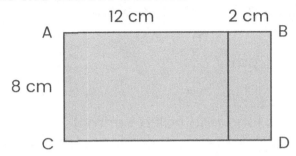

- (A) 8 cm × 12 cm × 2 cm
- (B) 8 cm + 12 cm × 8 cm + 2 cm
- (C) 8 cm × 12 cm + 8 cm × 2 cm
- (D) 8 cm × 14 cm

2. One-quarter of this garden is a patio.

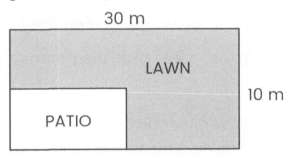

a. What is the area of the lawn? _____ m²
b. Patio tiles are 50 cm by 50 cm. How many tiles are needed to cover the patio? _____

3. If the area of a rectangle with a width equal to 4 m is the same as the perimeter of a triangle whose sides are of lengths 7 m, 8 m and 9 m. What is the perimeter of the rectangle?

- (A) 16 m
- (B) 18 m
- (C) 20 m
- (D) 22 m

GEOMETRIC MEASUREMENT

6.4 Perimeter and Area of Rectangles

4 Dr. Eric has a rectangular desk whose top is 30 cm wide and 40 cm long. Find the perimeter and the area of the top of the desk.

Perimeter = _____ cm

Area = _____ cm^2

5 A rectangular room is 8 m long and has a perimeter of 28 m. What is the area of the room? Square carpet tiles are 40 cm long. How many carpet tiles are needed to cover the floor of the room?

6 Which has a greater perimeter, a square with a side of 5 cm or a rectangle with a length of 6 cm and width 5 cm?

Ⓐ Perimeter of a square is greater than the perimeter of a rectangle.

Ⓑ The perimeter of a square is less than the perimeter of a rectangle.

7 Lucas wants to make a rectangular picture frame. The width of the photo is 4 cm while its length is 8 cm. Find the length of the framing material Lucas will need _____

8  The perimeter of a rectangular flowerbed is 180 cm. The area is 1400 cm^2. What are the dimensions of the flower bed?

The dimensions of the flower bed are _____ and _____ .

GEOMETRIC MEASUREMENT

Perimeter and Area of Rectangles 6.4

9. A rectangular living room has an area of 30 square meters and a perimeter of 26 meters. What are the dimensions of the living room?
   The dimensions of the flower bed are _____ and _____.

10. Mr. Caleb is planning to make a flower garden outside our classroom window. He wants the garden to be 18 feet long and 2 feet wide. If he decides to fence in this flower garden, how much fencing would Mr. Caleb need?

    A) 20 ft   B) 40 ft   C) 36 ft   D) 18 ft

11. A floor 6 m by 20 m is covered by square tiles with sides of 20 cm. How many of the small triangular tiles will there be?

    A) 10000   B) 12000
    C) 1240   D) 14400

12.
    A rectangular swimming pool with a length of 20 meters and width 9 meters are surrounded by grass. The path has a width of 5 meters. What will be the area of the path?

    A) 390 m²   B) 350 m²   C) 300 m²   D) 400 m²

GEOMETRIC MEASUREMENT

6.4 Perimeter and Area of Rectangles

13 Ryan wants to build a sidewalk around a garden that is 26 meters long and 16 meters wide. He wants the path to be 4 meters wide. What is the area of the sidewalk in square meters?

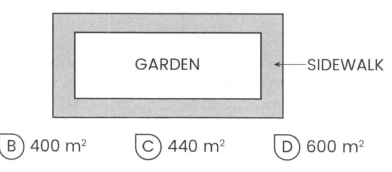

- A) 4500 m²
- B) 400 m²
- C) 440 m²
- D) 600 m²

14 The width of a painting is 8 in. and its length is 20 in. The painting has a frame around it. The frame is uniform with width 4 in.

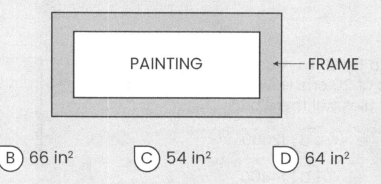

- A) 110 in²
- B) 66 in²
- C) 54 in²
- D) 64 in²

15 Dylan wants to draw the border lines of the basketball court that is 68 feet long and 40 feet wide. It takes Dylan 4 minutes to paint 6 feet. How long would it take Dylan to finish painting the borderlines?

GEOMETRIC MEASUREMENT

6.5 Chapter Review

1. The area of a rectangle is 400 cm². Its length is 25 cm. Find its width and perimeter.

2. The perimeter of a rectangle is 26 cm. Its length is 8 cm. Find its width and area.

3. A rectangle has an area of 10 square meters and a perimeter of 22 meters. What are the dimensions of the rectangle?
   - (A) 5 m × 2 m
   - (B) 10 m × 1 m
   - (C) 11 m × 2 m
   - (D) 11 m × 1 m

4. What is the area of the figure below?

   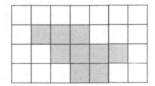

   1 square unit

   - (A) 16 square units
   - (B) 8 square units
   - (C) 9 square units
   - (D) 10 square units

GEOMETRIC MEASUREMENT

6.5 Chapter Review

5 What is the area of the shaded region?

Ⓐ 440 cm² Ⓑ 330 cm²
Ⓒ 420 cm² Ⓓ 320 cm²

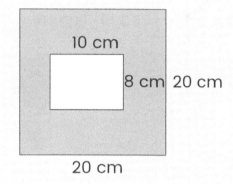

6 A rectangular walkway has an area of 220 m². What is the length of the walkway if the width is 11 m?

Ⓐ 20 m   Ⓑ 30 m   Ⓒ 40 m   Ⓓ 50 m

7 David uses paper clips to measure two sides of a rectangle.

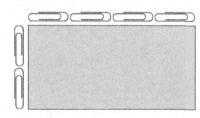

Select the two number sentences that can be used to calculate the number of paper clips David will need to go all the way around the rectangle.

Ⓐ 4+4+4+4   Ⓑ 4+2+4+2   Ⓒ 2×4   Ⓓ (4×2)+(2×2)   Ⓔ 4×2

GEOMETRIC MEASUREMENT

Chapter Review 6.5

8 The perimeter of a rectangular field is 90 m and its length is 30 m. What are the dimensions of the field?

(A) 20 m by 15 m (B) 30 m by 10 m
(C) 20 m by 10 m (D) 30 m by 15 m

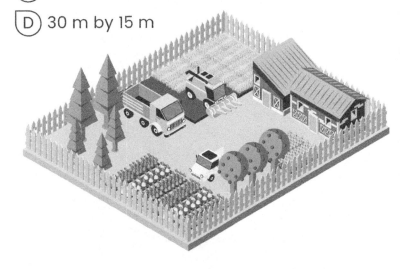

9 What is the perimeter of the polygon below if each side is equal to 7 inches?

(A) 84 inches (B) 74 inches
(C) 88 inches (D) 77 inches

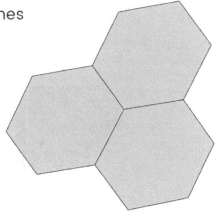

GEOMETRIC MEASUREMENT

6.5 Chapter Review

10 The shaded area of the picture below shows the portion of the wall Luke painted in 20 minutes.

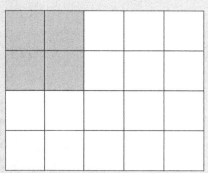

Working at the same pace, how long will it take Luke to paint the entire wall?

- A) 120 minutes
- B) 48 minutes
- C) 96 minutes
- D) 100 minutes

11 What is the perimeter of the figure below?

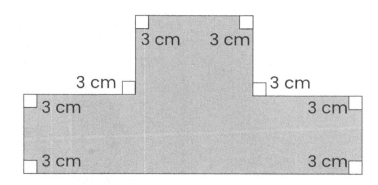

- A) 30 cm
- B) 21 cm
- C) 32 cm
- D) 28 cm

GEOMETRIC MEASUREMENT

Chapter Review 6.5

12 What is the area of the figure below?

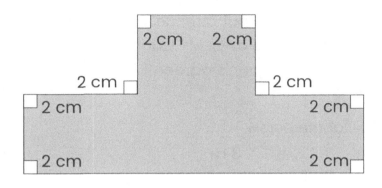

Ⓐ 20 cm²   Ⓑ 18 cm²   Ⓒ 16 cm²   Ⓓ 14 cm²

13 The diagram below shows four rectangles.

Rectangle 1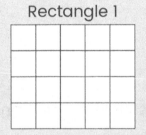

Rectangle 2

Rectangle 3

Rectangle 3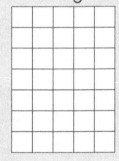

Which two rectangles have the same area?

_____

GEOMETRIC MEASUREMENT

6.5 Chapter Review

14 Mr. Isaac has a rectangular garden. The length of his garden is 16 yards and its width is 6 yards. He wants to put a fence around his garden.
a) How many yards of fence does Mr. Isaac need? _____
b) What is the area of Mr. Isaac's garden? _____

15 Consider the rectangle below.

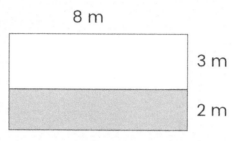

Which number sentence can be used to find the perimeter of rectangle ABCD?

Ⓐ (8+3+2+8+3+2) cm
Ⓑ 8×3×2 cm
Ⓒ 8×3+2 cm
Ⓓ 8+3+2 cm

16 Landon plants flowers in his yard. The width of the yard is 24 feet. The plan of the yard is shown below.

If the area of the yard is 272 square feet, what is the value of x, a foot?

_____

GEOMETRIC MEASUREMENT

Chapter Review 6.5

17 Consider the rectangle below.

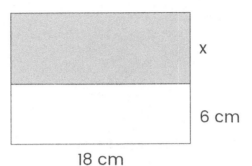

If the perimeter of the rectangle is 104 cm. what is the value of x, in cm?

18 Consider the rectangle below.

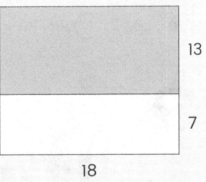

Which equation can be used to find the area of the rectangle?

A) $18 \times 7 \times 13$
B) $18 \times (13+7)$
C) $2 \times (18+7+13)$
D) $13 \times (18+7)$

GEOMETRIC MEASUREMENT

6.5 Chapter Review

19 A square is shown below.

The perimeter of the square can be represented by the expression 15x − 26. What is the side length of the square when x = 4?

Ⓐ 10 Ⓑ 22 Ⓒ 24 Ⓓ 34

20 A factory is 60 m long and has an area of 1320 m². Its warehouse has the same area but is two times longer. What is the perimeter of the warehouse?

Ⓐ 199 m Ⓑ 192 m Ⓒ 262 m Ⓓ 272 m

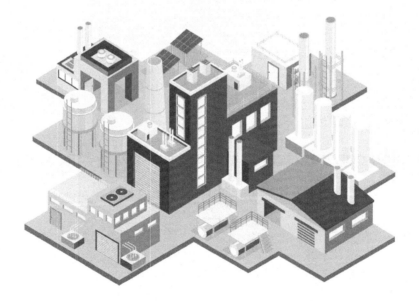

CHAPTER 7
REPRESENT AND INTERPRET DATA

REPRESENT AND INTERPRET DATA

7.1 Bar Graphs and Frequency Tables

Bar Graphs and Frequency Tables

Bar graphs

A bar graph represents data using columns placed in a 2-axes system.

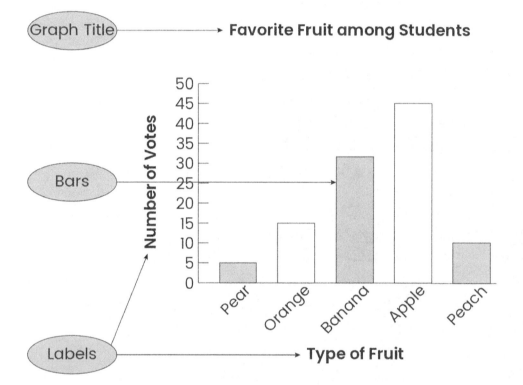

REPRESENT AND INTERPRET DATA

Bar Graphs and Frequency Tables

Example:
The table below shows the results of a survey asking grade 3 students what their favorite color is.

Favorite Color of Third Grade Students

Favorite Color	Tally	Totals
Red	|||| |||| |||| |||| ||	22
Yellow	|||| |||| |||	14
Blue	|||| |||| |||| |||| ||	22
Green	|||| |||| |||	13

We draw a vertical line and a horizontal line, one representing the different colors, and the other representing the number of votes. For each color, we draw a bar that reaches the same number on the number line as the number of votes. This is a bar graph.

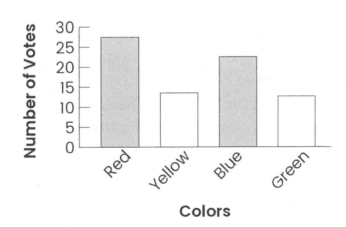

181

REPRESENT AND INTERPRET DATA

Bar Graphs and Frequency Tables

We must always give your graph a title and label on the vertical and horizontal lines. It is clear from the bar graph that red was the most popular color. Another way to draw a bar graph is:

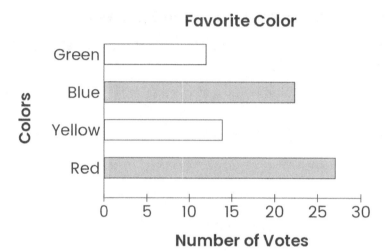

FREQUENCY TABLES

A frequency table is a visual summarization of data that shows the number of times a data value occurs. A frequency table has 3 columns with the headings category (the groups of data being evaluated), tally, and frequency.

For example, let's assume we are analyzing the spelling test scores of a 4th grade class. The scores of the class are 89, 83, 78, 90, 95, 95, 78, 78, 89, and 83. For our frequency table, we first determine the category we are analyzing which is test scores. Then we enter the test scores in ascending order into the frequency table. Then, we put a tally in the tally column for each time a score appears. Finally, we count the number of tallies for each score and write the quantity in the frequency column. We translate these test scores into the following frequency table:

REPRESENT AND INTERPRET DATA

Bar Graphs and Frequency Tables

Bar graph is always give your graph a title and label the vertical and horizontal lines. It is clear from the bar graph that red was the most popular color. Another way to draw a bar graph is:

Score	Tally	Frequency
78	III	3
83	II	2
89	II	2
90	I	1
95	II	2

Now that we have a frequency table, we can use it to answer questions about the headings category.

Example:
How many students took the math test?
To do this, we sum the numbers in the frequency column of the table.
3+2+2+1+2=10
Since the sum of the frequency column is 10 and each score is tallied only once, we know that 10 students took the math test.

Example:
What fraction of the students scored an 83 on the test?
We know that the total number of students that took the test is 10. Therefore, the denominator of a fraction is 10. According to the frequency table, the number of students that scored an 83 on the test is 2. This is our numerator.
Therefore, the fraction of students that scored an 83 on the test is $\frac{2}{10}$

REPRESENT AND INTERPRET DATA

7.1 Bar Graphs and Frequency Tables

1. The graph below shows the number of 4th grade students that chose their favorite types of fruit.

   Which of the following statements is true?

   A) Most students chose apple or cherry
   B) Most students chose kiwi or cherry
   C) Most students chose apple or orange
   D) None of the above is true

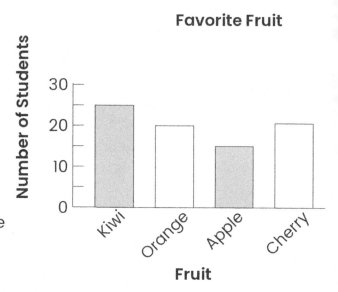

2. A group of students took a math test, and their scores were: 83, 96, 89, 76, 82, 84, 78, 96, 86, and 89. Which frequency table correctly represents the test scores?

A)

Score	Tally	Frequency
70 - 79	||	2
80 - 89	||||||	6
90 - 99	||	2

B)

Score	Tally	Frequency
70 - 79	||	2
80 - 89	|||||	5
90 - 99	||	2

REPRESENT AND INTERPRET DATA

Bar Graphs and Frequency Tables — 7.1

C)

Score	Tally	Frequency
70 - 79	||	2
80 - 89	|||||	5
90 - 99			1

D)

Score	Tally	Frequency
70 - 79	||	2
80 - 89	||||||	6
90 - 99			1

3 What was the highest recorded temperature?

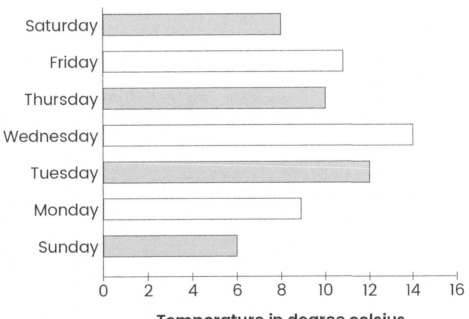

Recorded Temperature During One Week

A) 10 °C   B) 12 °C   C) 14 °C   D) 16 °C

REPRESENT AND INTERPRET DATA

7.1 Bar Graphs and Frequency Tables

4 There are 10 kids on a baseball team. Which frequency table correctly represents the number of home runs hit by the team if the total home runs per player are: 2, 16, 7, 16, 13, 4, 10, 6, 14, 15, 16, 17 and 18?

A

Homerun Range	Tally	Frequency
1 – 5	||	2
6 – 10	||	2
11 – 15	|||	3
16 – 20	||||	4

B

Homerun Range	Tally	Frequency
1 – 5	||	2
6 – 10	|||	3
11 – 15	|||	3
16 – 20	||||	4

C

Homerun Range	Tally	Frequency
1 – 5	||	2
6 – 10	|||	3
11 – 15	|||	3
16 – 20	|||||	5

D

Homerun Range	Tally	Frequency
1 – 5	||	2
6 – 10	||	2
11 – 15	|||	3
16 – 20	|||||	5

REPRESENT AND INTERPRET DATA

Bar Graphs and Frequency Tables — 7.1

5 Use the bar graph: How many people chose fish as their favorite meal?

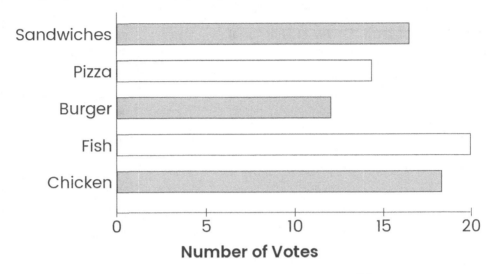

A) 5   B) 10   C) 15   D) 20

6 The table shows the amount of time the students in band spend practicing their instruments on a given evening.

How many students practiced for 30 minutes or less?

A) 20   B) 21
C) 17   D) 19

Time Spent Practicing	Tally	Frequency
1 – 10	|||| |||| ||	12
11 – 20	||||	4
21 – 30	||||	5
31 – 40	|||| ||	7
41 – 50	|||| |||	8
51 – 60	||	2

REPRESENT AND INTERPRET DATA

7.1 Bar Graphs and Frequency Tables

7. Seventy students were asked about their favorite types of fruit. Draw a bar graph using the below chart.

Favorite Fruits	
Grapes	IIII IIII IIII IIII
Oranges	IIII IIII IIII
Dragon Fruits	IIII IIII II
Pears	IIII IIII IIII II
Others	IIII I

8.

Pet	Tally	Frequency
Dog	IIII IIII II	12
Cat	IIII IIII	10
Fish	IIII II	7
Bird	IIII III	8
None	IIII III	8

The table shows the types of pets that students own in a 4th grade class.

How many more students have a dog or cat than a bird or fish?

Ⓐ 6   Ⓑ 7
Ⓒ 9   Ⓓ 10

REPRESENT AND INTERPRET DATA

Bar Graphs and Frequency Tables — 7.1

9. Grace invited her classmates to her birthday party. She passed a paper around asking everyone attending to write his/her choice for a meal. The paper she got back looked like this:

Burger	Pizza	Burger	Sandwich	Burger
Pizza	Pizza	Sandwich	Burger	Pizza
Pizza	Burger	Sandwich	Pizza	Burger
Sandwich	Sandwich	Burger	Sandwich	Burger
Burger	Sandwich	Sandwich	Burger	Pizza

Represent the data using a bar graph. Be sure to include a title and to label each axis.

_____

_____

10. The table below shows the number of bugs caught by a group of students over the course of a week.

   How many more bugs did the group catch on Thursday, Friday and Saturday than Sunday and Monday?

   A) 5
   B) 2
   C) 4
   D) 3

Day	Tally	Frequency
Sunday	|||||	5
Monday	||||	4
Tuesday	||	2
Wednesday	|||	3
Thursday	|||||	5
Friday	|||	3
Saturday	|||||	5

REPRESENT AND INTERPRET DATA

7.1 Bar Graphs and Frequency Tables

11 The chart below shows the average scores in Art, English, Math, and Science, for students in grades 4 and 5.

In which subject is the difference of averages between grade 4 and grade 5 equal to 20 points?

A) Math
B) Art
C) English
D) Science

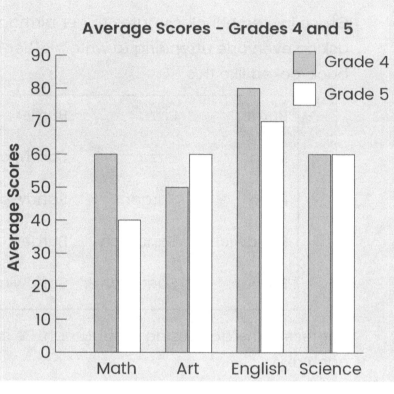

12

Month	Tally	Frequency
August	II	2
September	IIII	4
October	IIII III	8
November	II	2
December	III	3
January	III	3
February	III	3
March	IIII	4
April	II	2
May	IIII III	8

The table shows the number of absent students for each month of the school year.

What fraction of the absences occurred in December, January, February?

A) $\frac{5}{31}$
B) $\frac{3}{31}$
C) $\frac{6}{31}$
D) $\frac{9}{31}$

REPRESENT AND INTERPRET DATA

Bar Graphs and Frequency Tables — 7.1

13 A total of 100 students attended four different school events. The graph below shows the number of boys and girls attending each of the events.

Which event was attended by as many boys as girls?

A) Dance
B) Soccer
C) Art
D) Theater

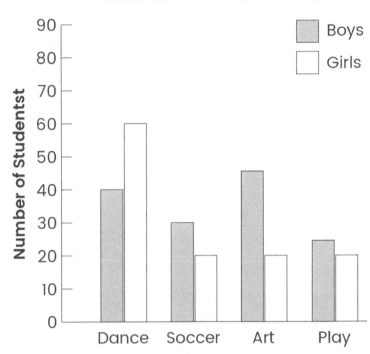

14

Year/Grade	Tally	Frequency
Freshman	𝍫𝍫 ||	12
Sophomore	𝍫𝍫𝍫𝍫	20
Junior	𝍫 |||	8
Senior	𝍫𝍫𝍫𝍫 ||	22

The table below shows the number of students participating in math club by grade.

What fraction of the math club students are freshmen and sophomores?

A) $\dfrac{30}{64}$
B) $\dfrac{32}{62}$
C) $\dfrac{42}{64}$
D) $\dfrac{28}{64}$

REPRESENT AND INTERPRET DATA

7.1 Bar Graphs and Frequency Tables

15 A survey was conducted to find out the favorite color among a group of students.

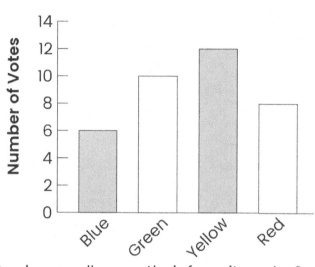

How many students chose yellow as their favorite color?

Ⓐ 6   Ⓑ 10

Ⓒ 12 Ⓓ 8

NEXT CHAPTER:
7.2 Line Plots and Categorical vs. Numerical Data

REPRESENT AND INTERPRET DATA

7.2 Line Plots and Categorical vs. Numerical Data

Line Plots and Categorical vs. Numerical Data

A line plot is a visual representation of data. To create a line plot, we draw a horizontal line with tick marks indicating all possible values. Then, we place an 'X' above each value as it corresponds to the data.

Once we have a line plot, we can answer questions about the data represented in the plot.

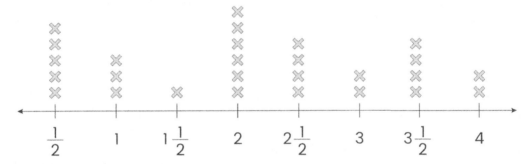

Let's look at a couple of example questions using the line plot showing the number of strawberries in the lunches of students in a 4th-grade students.

Example :

How many students had 2 or fewer strawberries in their lunch?

The possible values on the line plot that are 2 or less $\frac{1}{2}$, 1, $1\frac{1}{2}$, 1 and 2. To answer this question, we must add up the number of x's above each of the possible values that are 2 or less.

$$5+3+1+6=15$$

By adding the number of X's above each of the possible values, we determine that 15 students had 2 or less strawberries in their lunch.

REPRESENT AND INTERPRET DATA

Line Plots and Categorical vs. Numerical Data

CATEGORICAL VS. NUMERICAL DATA

Categorical data is often referred to as qualitative data. This type of data is often described as "words." Categorical data is often categorized by characteristics such as gender, hair or eye color, school, etc. Categorical data is often represented in a bar graph or a line plot. The bar graph shows the categorical data of how many birthdays fall into each category in a certain classroom.

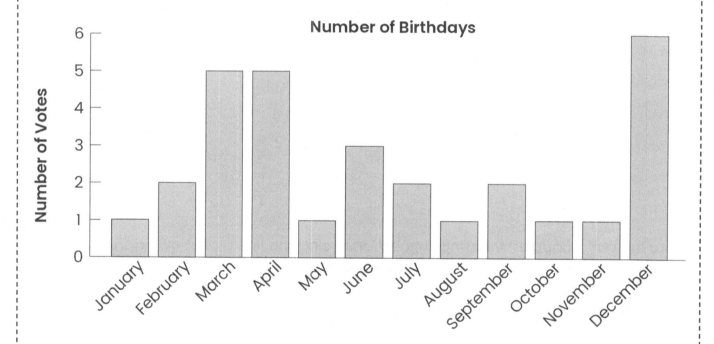

Numerical data is often referred to as quantitative data. This type of data is often described as numerical values. Numerical data is measurable quantities. These quantities may be miles run, age, height, shoe size, etc. Quantitative data is best represented in a line plot. The line plot below shows the numerical data of how many miles a group of students ran.

REPRESENT AND INTERPRET DATA

Line Plots and Categorical vs. Numerical Data

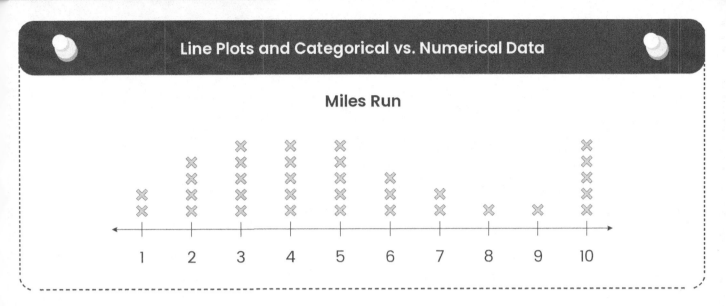

REPRESENT AND INTERPRET DATA

7.2 Line Plots and Categorical vs. Numerical Data

1 Use the data in the table to finish the line plot below.

Length of Pencils (cm)	
Pencil Length	Number of Pencils
$6\frac{1}{2}$	8
7	5
$7\frac{1}{2}$	2
9	4

2 Use the data in the table to finish the line plot below.

Baking Cookies	
Number of Batches Baked	Number of Bakers
$\frac{1}{2}$	2
$\frac{3}{4}$	4
1	5
$1\frac{1}{4}$	2
$1\frac{1}{2}$	1
$1\frac{3}{4}$	6
2	5

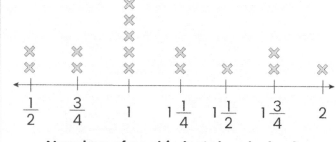

Number of cookie batches baked

REPRESENT AND INTERPRET DATA

Line Plots and Categorical vs. Numerical Data — 7.2

3 Select the line plot that accurately represents the data in the following table:

Time Spent Studying (in hrs)	Number of Students
$\frac{1}{4}$	2
$\frac{1}{2}$	1
$\frac{3}{4}$	4
1	5
$1\frac{1}{4}$	4
$1\frac{1}{2}$	3
$1\frac{3}{4}$	2
2	3

(A)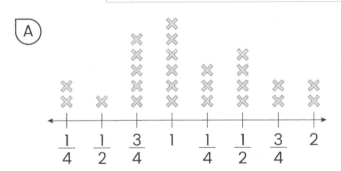

Time Spent Studying in Hours

(B)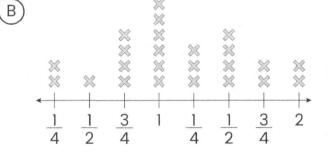

Time Spent Studying in Hours

197

REPRESENT AND INTERPRET DATA

7.2 Line Plots and Categorical vs. Numerical Data

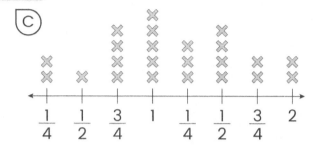

C

Time Spent Studying in Hours

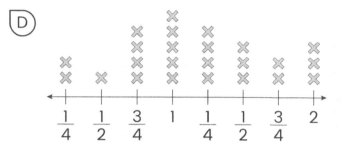

D

Time Spent Studying in Hours

4 Is the length of a leaf numerical or categorical data?

(A) Numerical

(B) Categorical

5 Ribbons of different colors were cut to decorate the hall.
Use the data in the table to create a line plot.

Length of the ribbon (in m)	Number of Ribbons
$\frac{1}{10}$	6
$\frac{2}{10}$	4
$\frac{3}{10}$	2
$\frac{4}{10}$	3
$\frac{5}{10}$	5
$\frac{6}{10}$	3

REPRESENT AND INTERPRET DATA

Line Plots and Categorical vs. Numerical Data — 7.2

6 Is gender numerical or categorical data?

(A) Numerical (B) Categorical

7 Is height numerical or categorical data?

(A) Numerical (B) Categorical

8 Is the amount of wheat flour used in recipes numerical or categorical data?

(A) Numerical (B) Categorical

9 Is shampoo flavor numerical or categorical data?

(A) Numerical (B) Categorical

10 The following figure shows the amount of water left in the water bottles of several students.

How many students had $\frac{4}{10}$ of a bottle (or less) of water left?

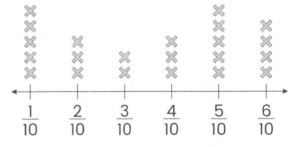

Amount of water left in the water bottles

(A) 10 (B) 11 (C) 12 (D) 13

REPRESENT AND INTERPRET DATA

7.2 Line Plots and Categorical vs. Numerical Data

11 Students in Ms. Helen's class selected orange juice cups based on how many ounces were in each cup. She counted the number of cups of orange juice her fourth-grade students selected. She presented the data on this line plot.

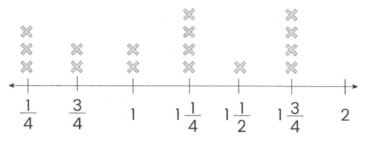

Orange juice in cups

How many students had less than 1 cup of orange juice?

---

12 Use the data in this table to complete the line plot.

Hours Studied				
Number of Hours Studied	$\frac{3}{4}$	1	$1\frac{1}{4}$	$1\frac{1}{2}$
Number of Students	4	6	3	1

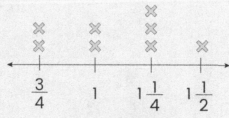

Hours Studied

200

REPRESENT AND INTERPRET DATA

Line Plots and Categorical vs. Numerical Data 7.2

13 Use this list of data to complete the line plot.

Hours of exercise: $\frac{1}{2}$, $1\frac{3}{4}$, $\frac{3}{4}$, $\frac{1}{2}$, $1\frac{3}{4}$, 1, 1, $\frac{1}{2}$, $1\frac{1}{4}$, $\frac{1}{2}$, $1\frac{1}{2}$, $\frac{1}{2}$, $1\frac{1}{4}$, $1\frac{3}{4}$

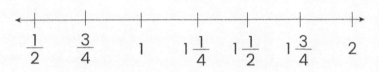

Hours of Exercise

14 Use the data in the table to complete the line plot

Length of Worms					
Worm Length (inches)	$2\frac{3}{4}$	3	$3\frac{1}{4}$	$3\frac{1}{2}$	4
Number of Worms	2	3	2	4	2

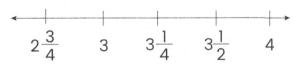

Length of Worms

REPRESENT AND INTERPRET DATA

7.2 Line Plots and Categorical vs. Numerical Data

15 Use this list of data to complete the line plot.

Cups of Sugar per Day: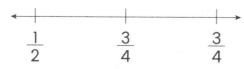

Cups of Sugar Per Day

NEXT CHAPTER:
7.3 Chater Review

REPRESENT AND INTERPRET DATA

7.3 Chapter Review

1. This line plot shows the amount of fuel, in gallons, used for 11 campsite grills. How many gallons of fuel do most of the grills use?

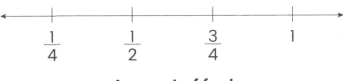

Amount of fuel

2. This line plot shows the length of Mr. Logan's pencils. How many pencils are longer than $8\frac{1}{4}$ inches?

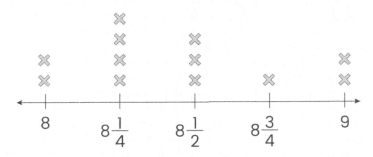

Pencil Length (in inches)

REPRESENT AND INTERPRET DATA

7.3 Chapter Review

3 This line plot shows the amount of paint, in cups, 8 people used in art class. How many people used less than $\frac{3}{4}$ cups of white paint?

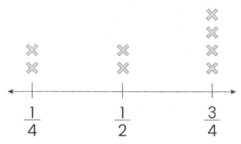

Amounts of White Paint

4 This line plot shows the amount of paint, in cups, 7 people used in art class. How many people used less than 1 cup of orange paint?

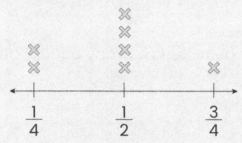

Amount of Orange Paint (in cups)

REPRESENT AND INTERPRET DATA

Chapter Review 7.3

5

Palette	Tablespoons
Palette 1	1
Palette 2	$\frac{3}{4}$
Palette 3	$\frac{3}{4}$
Palette 4	$\frac{3}{4}$
Palette 5	$\frac{2}{4}$
Palette 6	1
Palette 7	$\frac{2}{4}$
Palette 8	$\frac{3}{4}$
Palette 9	$\frac{3}{4}$

This table shows the amount of red paint on 9 palettes in an art class. Create a line plot from this data.

⟵――――――――――⟶
Amount of Red Paint (in tablespoons)

6 Mrs. Kayla volunteered to cook all the meals for the faculty on their camping retreat. She calculated how much propane fuel she would use during each meal and entered the amounts in the table below. Create a line plot to represent the amount of propane fuel needed.

Meal	Number of Cans
Breakfast 1	$\frac{2}{4}$
Lunch 1	1
Dinner 1	1
Breakfast 2	$\frac{3}{4}$
Lunch 2	$\frac{2}{4}$
Dinner 2	$\frac{3}{4}$
Breakfast 3	1

⟵――――――――――⟶
Number of Fuel Cans

REPRESENT AND INTERPRET DATA

7.3 Chapter Review

7.

| Hot Chocolate Mix ||
Students	Tablespoons
David	$\frac{2}{4}$
Ava	$\frac{1}{4}$
Noah	$\frac{2}{4}$
Emily	$\frac{3}{4}$
Ryan	$\frac{2}{4}$
Max	1
Mia	$\frac{1}{4}$
Blake	1

In his cooking class, Mr. Jason gave each student some hot dark chocolate mix. Use this data to create a line plot.

Hot Dark Chocoloate Mix (in tablespoons)

8. Use the data in this table to create a line plot.

Ounces of Cologne								
2	1	$1\frac{1}{4}$	$1\frac{1}{2}$	$1\frac{1}{2}$	1	2	$1\frac{1}{2}$	2

REPRESENT AND INTERPRET DATA

Chapter Review 7.3

9 Use the data in this table to complete the line plot.

Length (inches)	3	$3\frac{1}{4}$	$3\frac{1}{2}$	4
Number of Butterflies	3	4	2	3

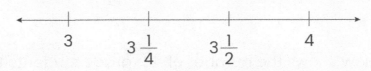

Butterfly Length (inches)

10 Use the data in this table to complete the line plot.

Button Used	$3\frac{3}{4}$	4	$4\frac{1}{4}$	$4\frac{1}{2}$	5
Number of People	5	6	4	2	3

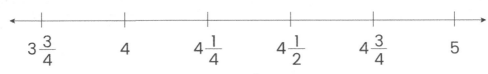

Amount of Buttons

REPRESENT AND INTERPRET DATA

7.3 Chapter Review

11 The following bar graph shows the categorical data of the favorite colors for a group of students.

How many more students like yellow or blue than red or pink?

(A) 3
(B) 4
(C) 5
(D) 6

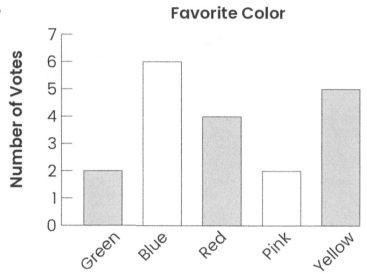

12 The graph below shows the number of 4th grade students that chose their favorite types of pets.

Which of the following statements is true?

(A) Most students chose cat or dog
(B) Most students chose dog or parrot
(C) Most students chose fish or dog
(D) None of the above is true

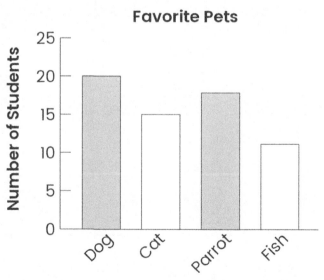

REPRESENT AND INTERPRET DATA

Chapter Review 7.3

13 The bar graph shows the qualitative data of fruit sold in a grocery store on a given morning.

How many fewer oranges than kiwis were sold?

Ⓐ 5 Ⓑ 6
Ⓒ 7 Ⓓ 8

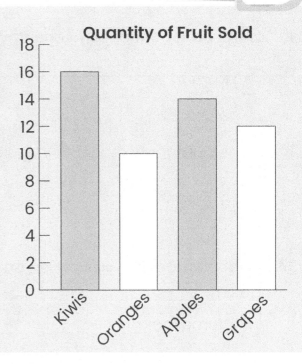

14 The line plot shows the numerical data of the number of bananas eaten by students over the course of a week. Each "X" indicates one student.

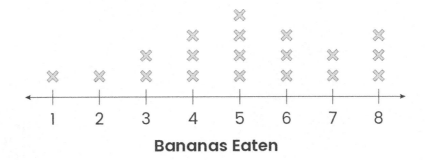

How many students ate an odd number of bananas?

Ⓐ 11 Ⓑ 9 Ⓒ 10 Ⓓ 8

REPRESENT AND INTERPRET DATA

7.3 Chapter Review

15 Is the brand of car someone drives numerical or categorical data?

　　Ⓐ Numerical　　　　　　　　Ⓑ Categorical

16 Is the number of students in a classroom numerical or categorical data?

　　Ⓐ Numerical　　　　　　　　Ⓑ Categorical

17 Is the number of miles driven on a daily commute numerical or categorical data?

　　Ⓐ Numerical　　　　　　　　Ⓑ Categorical

18 Are your favorite types of fruit numerical or categorical data?

　　Ⓐ Numerical　　　　　　　　Ⓑ Categorical

REPRESENT AND INTERPRET DATA

Chapter Review 7.3

19.

Month	Tally	Frequency
January	IIII	5
February	IIII III	8
March	IIII	4
April	II	2
May	III	3
June	II	2

The following frequency table shows the number of model airplanes sold at a hobby store each month for the first 6 months of the year. Use the table to answer the following questions.

What fraction of the model airplanes were sold in January and February?

A) $\frac{5}{3}$
B) $\frac{13}{22}$
C) $\frac{1}{3}$
D) $\frac{13}{24}$

20. The following line plot shows the length of leaves (in inches) measured by a group of students.

How many leaves with a length of more than $3\frac{3}{4}$ inches were measured?

A) 15
B) 16
C) 14
D) 20

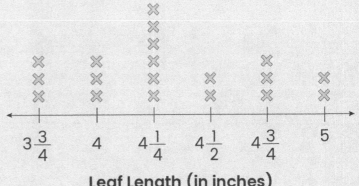

Leaf Length (in inches)

CHAPTER 8
MEASURING ANGLES

MEASURING ANGLES

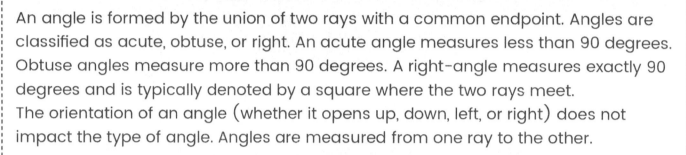

Types of Angles

An angle is formed by the union of two rays with a common endpoint. Angles are classified as acute, obtuse, or right. An acute angle measures less than 90 degrees. Obtuse angles measure more than 90 degrees. A right-angle measures exactly 90 degrees and is typically denoted by a square where the two rays meet.
The orientation of an angle (whether it opens up, down, left, or right) does not impact the type of angle. Angles are measured from one ray to the other.

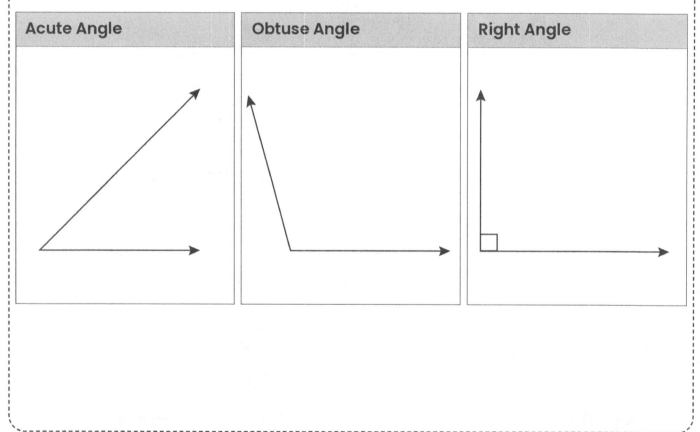

MEASURING ANGLES

Types of Angles 8.1

1 The minimum measurement of an angle is _____

(A) 0 (B) 90 (C) 180 (D) 360

2 **True or False:** These lines are perpendicular.

(A) True (B) False

3 **True or False:** 2 pairs of opposite sides are parallel.

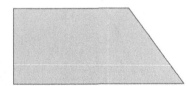

(A) True (B) False

4 An angle that measures less than 90 degrees is a(n) _____ angle.

(A) Acute (B) Obtuse (C) Right

215

MEASURING ANGLES

8.1 Types of Angles

5 **True or False:** Two pairs of opposite sides are parallel.

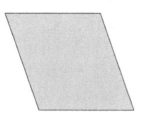

(A) True (B) False

6 An angle that measures exactly 90 degrees is a(n) _____ angle.

(A) Acute   (B) Obtuse   (C) Right

7 **True or False:** This triangle has 1 obtuse angle.

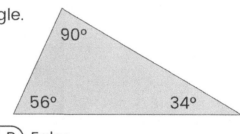

(A) True (B) False

8 Identify the following:

(A) Line   (B) Ray   (C) Point   (D) Edge

MEASURING ANGLES

Types of Angles 8.1

9. Andy connects Point M and Point N on this circle. What does he use to connect the two points?

A) Line B) Ray C) Line segment D) Edge

10. If two rays are 253 degrees to each other, what type of angle do they form?

A) Acute B) Obtuse C) Right

11. Peter draws this picture. What does he draw?

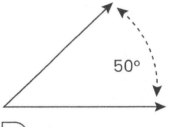

A) Line B) Point C) Angle D) Edge

12. Identify the following:

A) Line B) Ray C) Point D) Edge

MEASURING ANGLES

8.1 Types of Angles

13 If two rays are 35 degrees to each other, what type of angle do they form?

- (A) Acute
- (B) Obtuse
- (C) Right

14 Identify the following:

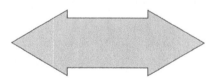

- (A) Line
- (B) Ray
- (C) Point
- (D) Edge

15 In the figure below, all angles that appear to be right angles are right angles. How many pairs of parallel line segments are in this shape?

- (A) 2
- (B) 3
- (C) 4
- (D) 5

NEXT CHAPTER:
8.2 Measuring Angles Using a Protractor

MEASURING ANGLES

8.2 Measuring Angles Using a Protractor

Measuring Angles Using a Protractor

A protractor is a tool used to measure an angle. A protractor has angle markings in 0 to 180 degrees. To make the protractor easier to read, angles that are multiples of 10 are written on the protractor and marked with the longest lines. Angles that are multiples of 5 are marked with a medium-sized line. Each individual angle marking is indicated with a short line. In this way, we can count 1, 5, or 10 to determine the measurement of an angle.

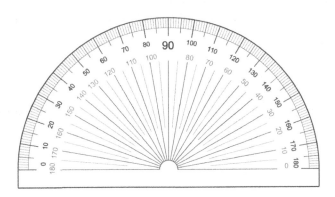

MEASURING ANGLES

Measuring Angles Using a Protractor

To measure an angle using a protractor, we line up one of the rays of the angle with the 0 degree marking. Then, we note where the other ray intersects with the edge of the protractor. For example, the following angle measures 42 degrees.

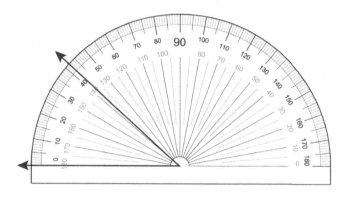

Additionally, we can measure an angle by counting the number of markings between the two rays of the angle. For example, the measurement of the following angle is 85 degrees.

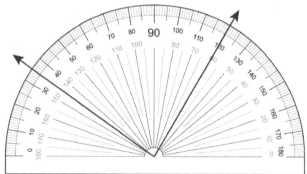

MEASURING ANGLES

Measuring Angles using a Protractor — 8.2

1. What is the measurement of the angle?

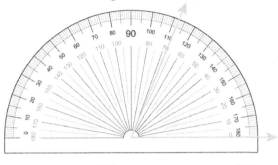

- (A) 111°
- (B) 68°
- (C) 88°
- (D) 98°

2. What is the measurement of the angle?

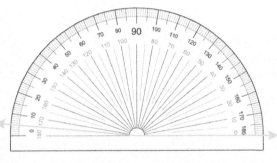

- (A) 11°
- (B) 8°
- (C) 90°
- (D) 175°

3. What type of angle is depicted by angle OB?

- (A) Acute
- (B) Obtuse
- (C) Right

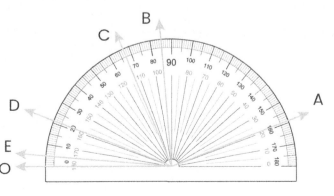

MEASURING ANGLES

8.2 Measuring Angles using a Protractor

4 What type of angle is depicted?

A) Acute   B) Obtuse   C) Right

5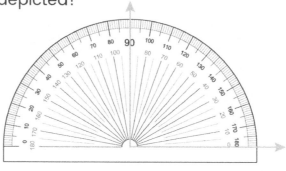

What is the measurement of angle OD?

A) 29°   B) 22°
C) 25°   D) 27°

6 What is the measure, in degrees, of this angle?

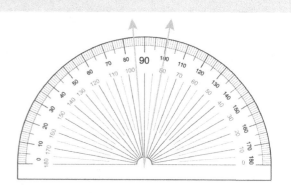

MEASURING ANGLES

Measuring Angles using a Protractor — 8.2

7 What is the measurement of this angle?

(A) 0°   (B) 120°   (C) 180°   (D) 270°

8 What is the measurement of this angle? Choose the best estimate.

(A) 360°   (B) 120°
(C) 180°   (D) 270°

9 Mia turns the dial on her machine 6 times. Each time she turns the dial, she moves it 30 degrees. What is the total number of degrees she turns the dial?

_____

10 The minute hand on a clock moves 1 degree every 60 seconds. How many degrees has the minute hand moved after 1200 seconds?

_____

MEASURING ANGLES

8.2 Measuring Angles using a Protractor

11 The blades of a fan move 180 degrees each second. How many seconds does it take for the blades to turn a complete circle?

12 What type of angle is depicted?

A) Acute B) Obtuse C) Right

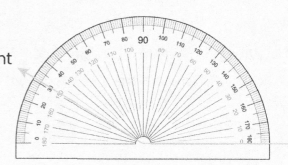

13 What is the measurement of this angle?

A) 360°   B) 120°   C) 180°   D) 270°

MEASURING ANGLES

Measuring Angles using a Protractor — 8.2

14 What is the measurement of the angle?

(A) 47° (B) 46°
(C) 78° (D) 34°

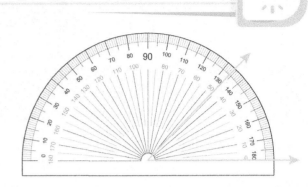

15

What is the measurement of angle EC?

(A) 22° (B) 15°
(C) 20° (D) 76°

NEXT CHAPTER:
8.3 Finding Unknown Angles

MEASURING ANGLES

8.3 Finding Unknown Angles

Finding Unknown Angles

Angles can be added together to make larger angles or split into pieces to make smaller angles. The measurements of two angles placed side by side with no overlap can be added together to find the measurement of the larger angle. Similarly, a large angle can be split into two smaller angles. We can use addition and subtraction to find the measurements of the angles.

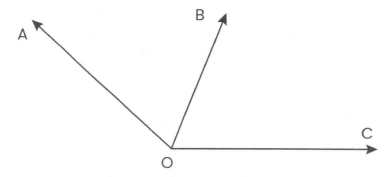

Example:

Let angle AOB measure 55 degrees and angle BOC measure 84 degrees. What is the measurement of AOC?

Since there is no overlap between angles AOB and BOC, we can add the measurements of their angles to find the measurement of angle AOC.

$$55+84=139$$

Therefore, angle AOC measures 139 degrees.

MEASURING ANGLES

Finding Unknown Angles — 8.3

1. A large angle is split into two smaller angles by a ray. The large angle measures 188 degrees and one of the smaller angles measures 166 degrees. What is the measurement of the other small angle?

   A) 20°   B) 22°   C) 24°   D) 26°

2. What is the measure of angle u?

   _____ °

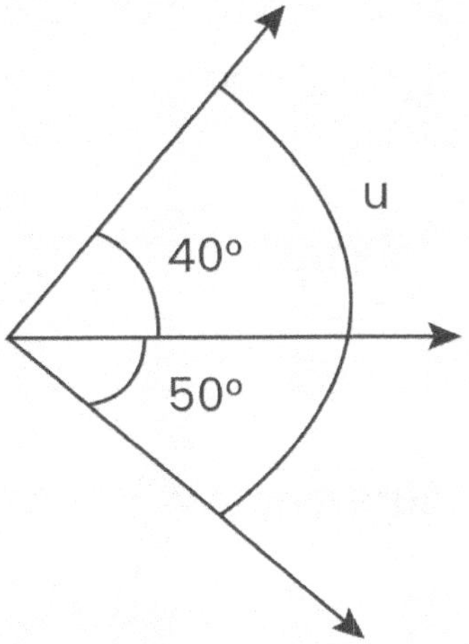

   (diagram shows angles of 40° and 50°, with angle u above)

3. What is the measure, in degrees, of angle z?

   (diagram shows angles 40.6° and 82.4°, with angle z)

   _____

MEASURING ANGLES

8.3 Finding Unknown Angles

4 A large angle is split into two smaller angles by a ray. One of the smaller angles measures 72 degrees and the other measures 44 degrees. What is the measurement of the large angle?

(A) 110°   (B) 100°   (C) 94°   (D) 116°

5 A large angle is split into two smaller angles by a ray. The large angle measures 134 degrees, and one of the smaller angles measures 88 degrees. What is the measurement of the other small angle?

(A) 48°   (B) 10°   (C) 46°   (D) 59°

6 What is the measure angle w?

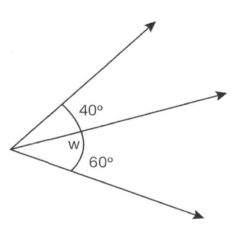

MEASURING ANGLES

Finding Unknown Angles — 8.3

7 What is the measure, in degrees, of angle x?

_____

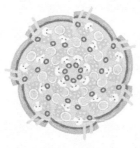

8 Parker and Chris are sharing a pizza cut into 7 large pieces. This model represents a pizza.

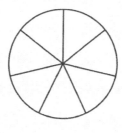

Parker and Chris each take 3 pieces of pizza and decide to divide the last piece equally. Which expression can be used to determine the angle created by dividing the last piece of pizza into two equal parts?

Ⓐ 360÷7

Ⓑ 360÷7+3

Ⓒ 360÷2×7

Ⓓ (360÷7)÷3

MEASURING ANGLES

8.3 Finding Unknown Angles

9. What is the measure, in degrees, of angle m?

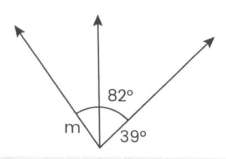

10.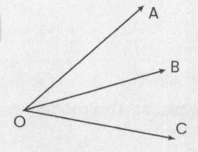
If angle AOB measures 15 degrees and angle BOC measures 4 degrees, what is the measurement of angle AOC?

11. If angle MON measures 52 degrees and angle PON measures 18 degrees, what is the measurement of angle MOP?

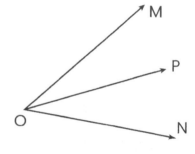

12. A large angle is split into two smaller angles by a ray. The large angle measures 111 degrees, and one of the smaller angles measures 22 degrees. What is the measurement of the other small angle?

   A) 122°   B) 133°   C) 66°   D) 89°

MEASURING ANGLES

Finding Unknown Angles — 8.3

13. A large angle is split into two smaller angles by a ray. One of the smaller angles measures 70 degrees, and the other measures 73 degrees. What is the measurement of the large angle?

Ⓐ 104° Ⓑ 153° Ⓒ 143° Ⓓ 3°

14. 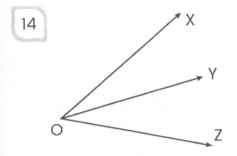 If angle XOZ measures 52 degrees and angle YOZ measures 40 degrees, what is the measurement of angle XOY?

15. If angle AOB measures 33 degrees and angle BOC measures 85 degrees, what is the measurement of angle AOC?

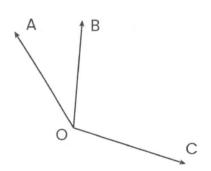

NEXT CHAPTER:
8.4 Chapter Review

MEASURING ANGLES

8.4 Chapter Review

1. What is the measure of this angle? Choose the best estimate.

   A) 20°
   B) 90°
   C) 140°
   D) 86°

2. How many 1-degree angles are in an angle that turns through a full circle?

3. How many 1-degree angles are in an angle that turns through a half-circle?

4. Emily draws an angle twice as large as the measure of the angle shown. What is the measure of Emily's angle?

   A) 56°
   B) 112°
   C) 100°
   D) 96°

   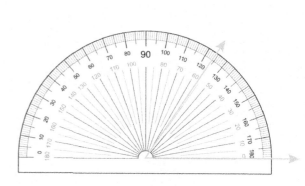

MEASURING ANGLES

Chapter Review 8.4

5 If angle AOB measures 22 degrees and angle BOC measures 72 degrees, what is the measurement of angle AOC?

_____ °

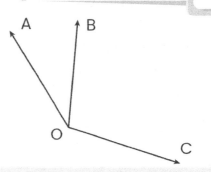

6 What is the measure of Angle m?

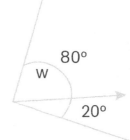

7 What is the measure of Angle w?

8 Mr. Peterson is building 2 new fences to separate his livestock. His property has the shape of a square.

What is the measure, in degrees, of the missing angle?

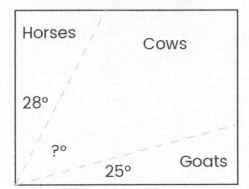

Mr. Peterson's Farm

MEASURING ANGLES

1.5 Chapter Review

9 The measure of angle G is 40°. What fraction of a complete circle is the measure of angle G?

10 Which fraction of a complete circle is the best estimate for the measure of this angle?

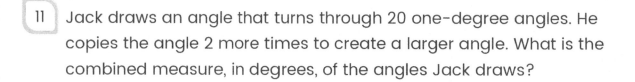

- Ⓐ $\dfrac{1}{360}$
- Ⓑ $\dfrac{50}{360}$
- Ⓒ $\dfrac{360}{20}$
- Ⓓ $\dfrac{40}{360}$

11 Jack draws an angle that turns through 20 one-degree angles. He copies the angle 2 more times to create a larger angle. What is the combined measure, in degrees, of the angles Jack draws?

12 The wheels on a bicycle turn 1 degree every $\dfrac{1}{360}$ seconds. How many degrees does the bicycle wheel turn in 3 seconds?

MEASURING ANGLES

Chapter Review 8.4

13 Ryan is drawing a 50 angle. Draw the second ray to create this angle.

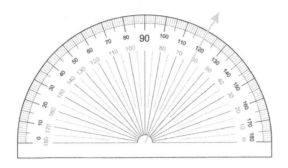

14 Write 2 expressions that determine the measure of this angle.

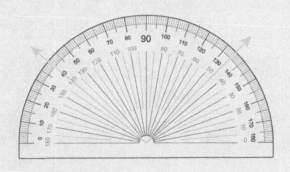

_____

15 The measure of angle WXZ is 120. John states that the measure of Angle ZXY is 52. Do you agree with John?

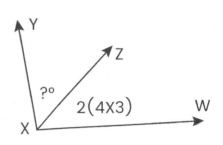

_____

MEASURING ANGLES

1.5 Chapter Review

16 Write the steps you would use to find the combined measure, in degrees, of these two angles.

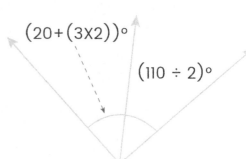

$(20+(3\times2))°$

$(110 \div 2)°$

_____

17 What types of angles are in this triangle?

87°
62°   36°

_____

18 What type of triangle is this?

90°
60°   30°

_____

MEASURING ANGLES

Chapter Review 8.4

19 What type of angle is shown?

_____

20 What type of angle is shown?

_____

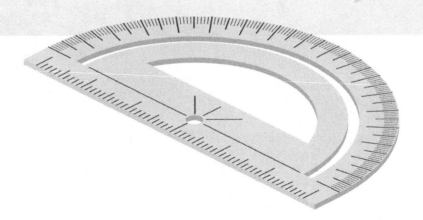

CHAPTER 9

GEOMETRY

GEOMETRY

9.1 Angles and Sides of Quadrilaterals and Triangles

Angles and Sides of Quadrilaterals and Triangles

We can classify quadrilaterals using several different approaches. Remember, a quadrilateral is a polygon with 4 sides, just like a triangle is a polygon with 3 sides. Triangles are broken into 3 categories: isosceles, right, and scalene. An isosceles triangle has two congruent sides, with the opposite angles also being congruent. A right triangle contains a right triangle. A scalene triangle is neither isosceles nor right. We can also describe triangles by the types of angles found within them. If all of the angles are congruent, then the triangle is equilateral (the angles are all congruent and the sides are all congruent). If all of the angles within the triangle are less than 90 degrees, then the triangle is acute. If one of the angles within the triangle is greater than 90 degrees, then the triangle is obtuse. Again, if one of the angles of the triangle is equal to 90 degrees, the triangle is a right triangle.

Acute Triangle	Obtuse Triangle	Right Triangle
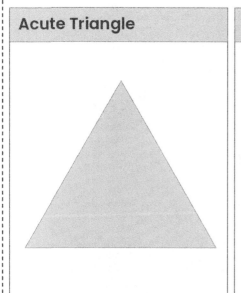		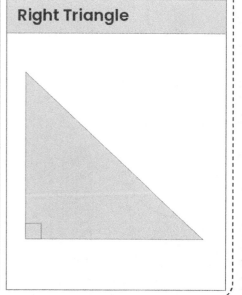

GEOMETRY

Angles and Sides of Quadrilaterals and Triangles

Isosceles Triangle	Equilateral Triangle	Scalene Triangle
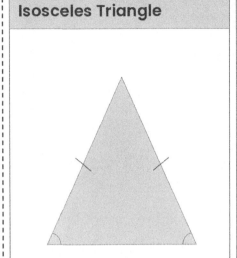	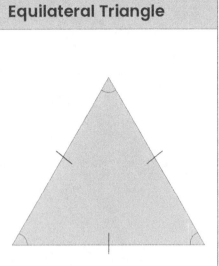	

There are several different types of quadrilaterals. These include parallelograms, squares, rectangles, rhombi, kites, and trapezoids.

Parallelogram	Square	Rectangle
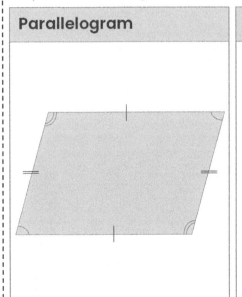		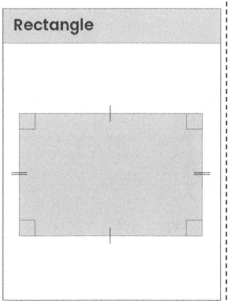

GEOMETRY

Angles and Sides of Quadrilaterals and Triangles

Rhombus

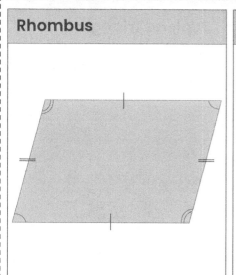

Kite

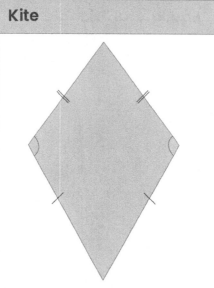

Trapezoid

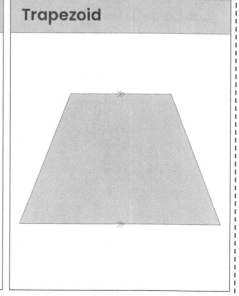

GEOMETRY

Angles and Sides of Quadrilaterals and Triangles — 9.1

1. What name best describes this shape?

   (A) Square (B) Trapezoid (C) Parallelogram (D) Kite

2. Lisa describes this triangle as an equilateral triangle. Angle describes this triangle as an isosceles triangle. Whose statement is correct?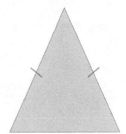

   (A) Lisa (B) Angel

3.  How many right angles does this quadrilateral have?

   (A) 4 (B) 2 (C) 3 (D) 6

4. What type of triangle is shown?

   (A) Right (B) Isosceles (C) Scalene (D) Equilateral

GEOMETRY

9.1 Angles and Sides of Quadrilaterals and Triangles

5 Classify the polygon.

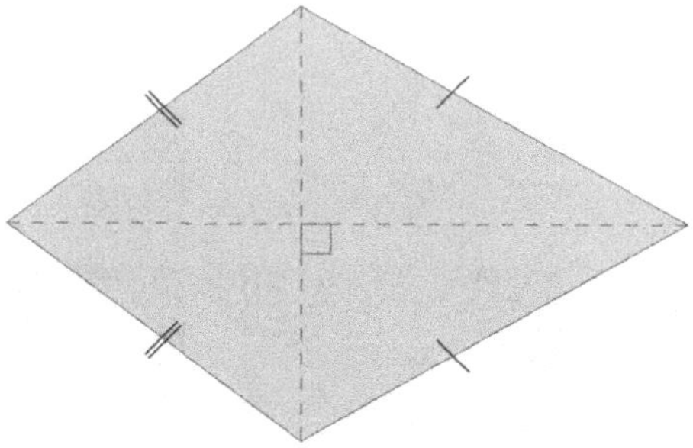

- (A) Kite
- (B) Trapezoid
- (C) Square
- (D) Rectangle

6 What type of triangle has only one angle that measures more than 90 degrees?

- (A) Scalene
- (B) Isosceles
- (C) Obtuse
- (D) Equilateral

7 Andrew drew a shape with 3 equal sides that are also equiangular. Which is the best description of the shape he drew?

- (A) Acute
- (B) Equilateral
- (C) Isosceles
- (D) Right

8 Which shape is a trapezoid?

A   B   C   D

9 What type of triangle has an interior angle of 90 degrees (a right angle)?

- (A) Isosceles
- (B) Equilateral
- (C) Right
- (D) Acute

GEOMETRY

Angles and Sides of Quadrilaterals and Triangles — 9.1

10 **True or False:** The name that best describes this shape is the quadrilateral.

- (A) True
- (B) False

11 Maria drew a shape with 3 sides and all three angles measure less than 90°. Which is the best description of the shape she drew?

- (A) Isosceles
- (B) Equilateral
- (C) Right
- (D) Acute

12 Fill in the blanks.

A triangle with one right angle and _____ acute angle(s) is called a Right angle triangle.

- (A) one
- (B) two
- (C) three
- (D) four

13 What name best describes this shape?

- (A) Pentagon
- (B) Trapezoid
- (C) Rhombus
- (D) Quadrilateral

GEOMETRY

9.1 Angles and Sides of Quadrilaterals and Triangles

14 Which two quadrilaterals have 4 congruent sides?

(A) Square and Rhombus (B) Square and Rectangle

(C) Rhombus and Rectangle (D) Rhombus and Kite

15 What is true of the angle bisectors of a kite?

(A) Obtuse (B) Parallel (C) Acute (D) Perpendicular

NEXT CHAPTER:
9.2 Parallel and Perpendicular Lines of Quadrilaterals and Triangles

9.2 Parallel and Perpendicular Lines of Quadrilaterals and Triangles

Parallel and Perpendicular Lines of Quadrilaterals and Triangles

A triangle is a polygon with 3 sides. The sum of the interior angles of a triangle is 180 degrees. Since they are 3-sided figures, triangles cannot have parallel lines. However, a right triangle does have a set of perpendicular lines. Recall, perpendicular lines form a right angle. As a right triangle has a right angle, the two sides of the triangle that form the right angle are perpendicular.

Quadrilaterals are polygons with 4 sides. The sum of the interior angles of a quadrilateral is 360 degrees. A parallelogram is a quadrilateral with opposite sides parallel and opposite angles congruent. A square has 4 congruent sides and 4 right angles. The adjacent sides of a square are perpendicular and the opposite sides of a square are parallel. Similarly, a rectangle has opposite congruent sides and 4 right angles. The adjacent sides of a rectangle are perpendicular and the opposite sides of a rectangle are parallel. A rhombus has 4 congruent sides and opposite congruent angles. Additionally, the opposite sides of a rhombus are parallel. A rhombus in which the adjacent sides are perpendicular is a square. A trapezoid is a quadrilateral in which one set of opposite sides are parallel. A kite has two pairs of adjacent congruent sides and one set of opposite congruent angles. Kites do not have parallel or perpendicular lines.

GEOMETRY

Parallel and Perpendicular Lines of Quadrilaterals and Triangles

However, the angle bisectors (a line from one angle to the opposite angle that divides both angles in half) are perpendicular.

Parallelogram	Square	Rectangle

Rhombus	Trapezoid	Kite

GEOMETRY

Parallel and Perpendicular Lines of Quadrilaterals and Triangles — 9.2

1. What name best describes this shape?

   A) Square   B) Trapezoid   C) Parallelogram   D) Kite

2. How many pairs of parallel lines does an obtuse triangle have?

   A) 3   B) 1   C) 2   D) 0

3. What type of triangle is shown?

   A) Equilateral   B) Isosceles   C) Right   D) Acute

4. **True or False:** A parallelogram has two pairs of perpendicular lines.

   A) True   B) False

5. Classify the polygon.

   A) Rectangle   B) Square
   C) Rhombus   D) Kite

GEOMETRY

9.2 Parallel and Perpendicular Lines of Quadrilaterals and Triangles

6 How many pairs of parallel lines must a quadrilateral have?

(A) 3   (B) 2   (C) 0   (D) 1

7 True or False: The name that best describes this shape is a kite.

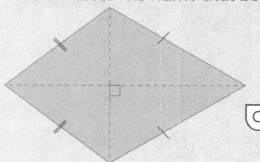

(C) True   (D) False

8 How many perpendicular lines does a square have?

(A) 3   (B) 4   (C) 2   (D) 1

9 What name best describes this shape?

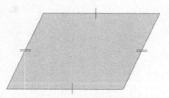

(A) Square   (B) Trapezoid   (C) Kite   (D) Parallelogram

GEOMETRY

Parallel and Perpendicular Lines of Quadrilaterals and Triangles — 9.2

10 **True or False:** A trapezoid has two pairs of parallel lines.

  (C) True   (D) False

11 Andrea draws a polygon. All of the interior angles are obtuse angles. Which shape did she draw?

(A) 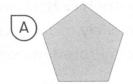   (B)   (C)    (D)

12 How many pairs of parallel lines exist in this polygon?

  (A) 2   (B) 3   (C) 4   (D) 1

13 **True or False:** A rectangle has 4 perpendicular lines.

  (C) True   (D) False

14 A rhombus with 4 perpendicular sides is a _____.

  (A) Rectangle   (B) Trapezoid   (C) Square   (D) Kite

15 The angle bisectors of a kite are perpendicular.

  (A) Always   (B) Sometimes   (C) Never

GEOMETRY

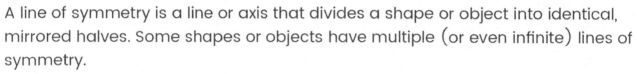

Lines of Symmetry

A line of symmetry is a line or axis that divides a shape or object into identical, mirrored halves. Some shapes or objects have multiple (or even infinite) lines of symmetry.

For example, a circle has an infinite number of lines of symmetry. Each dotted line in the figure below represents a line of symmetry. Additionally, infinitely many more lines of symmetry can be drawn to divide the circle into identical, mirrored halves.

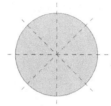

On the other hand, some figures have fewer lines of symmetry or no lines of symmetry. For example, the letter A shown below has only one line of symmetry. The second and third images are not lines of symmetry as folding the object along the dotted line does not result in identical, mirrored halves.

Line of Symmetry	Not a Line of Symmetry	Not a Line of Symmetry
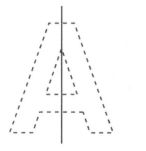	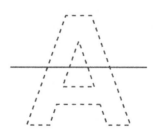	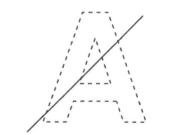

GEOMETRY

Lines of Symmetry 9.3

1. **True or False:** The dotted line in the image below is a line of symmetry.

   A) True   B) False

2.

   True or False: The dotted line in the image below is a line of symmetry.

   A) True   B) False

3. How many lines of symmetry does a scalene triangle have?

   A) 1   B) 2   C) 0   D) 3

4. **True or False:** The dotted line in the image below is a line of symmetry.

   A) True   B) False

GEOMETRY

9.3 Lines of Symmetry

5 **True or False:** The dotted line in the image below is a line of symmetry.

A) True   B) False

6 Which of the following capital letters have at least one line of symmetry?

A) F   B) N   C) M   D) S

7 **True or False:** The dotted line in the image below is a line of symmetry.

A) True   B) False

8 **True or False:** The dotted line in the image below is a line of symmetry.

A) True   B) False

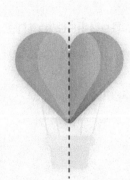

GEOMETRY

Lines of Symmetry 9.3

9. **True or False:** The letter "I" has 2 lines of symmetry.

   A) True B) False

10. Which of the following capital letters has at least one lines of symmetry?

    A) F B) N C) O D) S

11.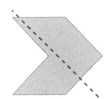

    True or False: The dotted line in the image below is a line of symmetry.

    A) True B) False

12. **True or False:** The dotted line in the image below is a line of symmetry.

    A) True B) False

    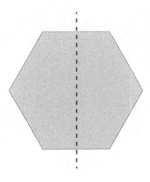

GEOMETRY

9.3 Lines of Symmetry

13 Which of the following capital letters has at least one lines of symmetry?

(A) F   (B) N   (C) X   (D) S

14 **True or False:** The letter "T" has 1 line of symmetry.

(A) True   (B) False

15 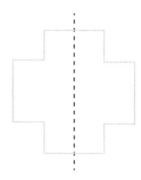 **True or False:** The dotted line in the image below is a line of symmetry.

(A) True   (B) False

GEOMETRY

9.4 Chapter Review

1. What type of triangle is shown?

   A) Acute
   B) Isosceles
   C) Equilateral
   D) Obtuse

2. How many pairs of parallel lines does an acute triangle have?

   A) 0
   B) 1
   C) 2
   D) 3

3.  **True or False:** The dotted line in the image below is a line of symmetry.

   A) True
   B) False

4. Elisa describes this shape as a parallelogram. Blessy describes this shape as a rhombus. Whose statement is correct?

   A) Blessy
   B) Elisa

GEOMETRY

9.4 Chapter Review

5 The dotted line through the figure is a line of symmetry.

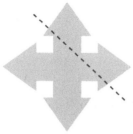

 A) Yes   B) No

6 What name best describes this shape?

A) Square   B) Trapezoid
C) Rhombus   D) Kite

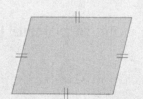

7 **True or False:** A trapezoid has 2 pairs of perpendicular lines.

A) True   B) False

8 Which shape is a kite?

 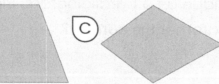

GEOMETRY

Chapter Review 9.4

9 How many lines of symmetry does an equilateral triangle have?

(A) 1 (B) 3 (C) 2 (D) Infinite

10 A rectangle is a parallelogram.

(A) Always (B) Sometimes (C) Never

11 **True or False:** This shape is an acute triangle.

(A) True (B) False

12 Which of the following capital letters have at least one line of symmetry?

(A) F (B) N (C) O (D) S

13 Fill in the blank.

A _____ triangle has no congruent sides and no congruent angles.

(A) Isosceles (B) Scalene (C) Right (D) Equilateral

14 How many lines of symmetry does a pentagon have?

(A) 5 (B) 2 (C) 0 (D) 3

GEOMETRY

9.4 Chapter Review

15 What type of polygon has exactly four sides?

- A) Triangle
- B) Pentagon
- C) Quadrilateral
- D) Hexagon

16 What is the sum of the interior angles of a triangle?

- A) 90
- B) 270
- C) 360
- D) 180

17 Which of the following capital letters has at least one line of symmetry?

- A) Z
- B) X
- C) D
- D) S

18 True or False: This quadrilateral is a trapezoid.

- A) True
- B) False

19 Which of the following capital letters has no line of symmetry?

- A) O
- B) N
- C) D
- D) M

20 What two dimensional figure has more than an infinite number of lines of symmetry?

COMPREHENSIVE ASSESSMENT - 1

COMPREHENSIVE ASSESSMENT

1 Comprehensive Assessment

1 Tom has 5 boxes of chocolates. Each box has 20 chocolates in it. So, Tom has 100 chocolates. Write a multiplication equation that matches this story.

2 Allen is 9 times as old as Adam. Adam is 6 years old. Write an equation to determine how old Allen is.

3 A coffee shop charges a cup of coffee for $7. The shop makes $4,200 selling coffee. How many cups of coffee does the shop sell?

(A) 200   (B) 600   (C) 700   (D) 550

4 Blake's mother told him to spend $\frac{5}{7}$ of an hour practicing basketball. Blake practiced basketball for $\frac{6}{8}$ of an hour. Did Blake practice the required amount of time? Explain your reasoning.

COMPREHENSIVE ASSESSMENT

Comprehensive Assessment 1

5. Which is more, 2,518 millimeters or 2 meters?

6. The dimensions of a math book is 12 inches tall and 8 inches wide. What is its area?

   A) 108 B) 72 C) 88 D) 96

7. David has 11 model motorbikes. Elvis has three times as many model motorbikes as David. How many more model motorbikes do they need to have altogether for a total of 60 model motorbikes?

   A) 10 B) 24 C) 16 D) 38

8. A train picked up 91 passengers at its third stop. This number is 13 times the number of passengers that it picked up at its second stop. How many passengers were picked up at the second stop?

   A) 7 B) 13 C) 23 D) 27

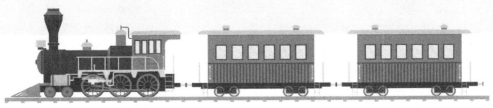

COMPREHENSIVE ASSESSMENT

1 Comprehensive Assessment

9 There are 15 tables in a restaurant. Each table can seat 6 customers. Tonight, 45 customers are eating dinner. How many empty seats are there?

10 Gabe travels 3,726 miles from his house to the beach. He makes two round trips on this route. How many miles does he travel?

(A) 10,126   (B) 14,904   (C) 16,642   (D) 12,864

11 Mark has 18 shoes. Six shoes are black, and the rest are brown. Choose the pair of fractions that shows equivalent fractions for the number of brown shoes in Mark's collection.

(A) $\frac{6}{9}$ and $\frac{4}{6}$ (B) $\frac{10}{9}$ and $\frac{5}{6}$ (C) $\frac{6}{10}$ and $\frac{4}{12}$ (D) $\frac{7}{9}$ and $\frac{4}{7}$

12 The thickness of a cent is $1\frac{7}{20}$ millimeters. A dollar is $\frac{90}{60}$ millimeters thicker than a cent. What is the thickness, in mm of a dollar?

COMPREHENSIVE ASSESSMENT

Comprehensive Assessment 1

13. A school has 25 classrooms. Each classroom has 5 windows. How many windows does the school have? Write an equation that matches the story. Use q to represent the unknown number.

14. There are 8 boxes of pens. Each box has 12 pens in it. If 10 students each receive the same number of pens, how many pens will be left over?

15. Is 6 a factor of 92?

16. A bag holds 104 pencils. Half of the pencils are red, and 25 pencils are green. The rest of the pencils are black. How many of the pencils are black? xplain your thinking.

COMPREHENSIVE ASSESSMENT

1   Comprehensive Assessment

17. Harvey is counting his shirts. Out of 21 shirts, 14 are white shirts, and the rest are black shirts. Which fraction is equivalent to the fraction of black shirts in Harvey's shirt collection?

(A) $\frac{1}{3}$ (B) $\frac{1}{9}$ (C) $\frac{6}{10}$ (D) $\frac{7}{9}$

18. A basket contains apples, oranges, mangoes, and pears. This table shows the amount of fruit represented as a fraction.

Fruits	Apples	Oranges	Mangoes	Pears
Fraction of basket	$\frac{10}{20}$	$\frac{60}{200}$	$\frac{20}{200}$	$\frac{1}{20}$

Kingsley says most of the basket has oranges and mangoes.
Do you agree with Kingsley? Explain your reasoning.

_____

_____

19. A square has 4 sides. Altogether how many sides do 17 squares have?

(A) 52   (B) 68   (C) 49   (D) 76

20. Gil bought 522 lemons and put them in boxes. There are 16 lemons in each box. How many boxes does Gil use?

(A) 49   (B) 42   (C) 31   (D) 33

COMPREHENSIVE ASSESSMENT

Comprehensive Assessment 1

21 There are 81 hamsters and 9 rats in the pet shop. How many times more hamsters are there than rats?

(A) 8   (B) 12   (C) 9   (D) 15

22 Mariah bought 6 boxes of erasers. There were 24 erasers in each box. How many erasers did Mariah buy? Write an equation to solve where y equals the number of erasers.

(A) 6×24=y
y=144

(B) 24−6=y
y=14

(C) 6+24=y
y=40

(D) 24÷6=y
y=2

23 What number has 6 ten thousands, 2 fewer thousands than ten thousands, 4 more hundreds than thousands, 1 more tens than hundreds, and 2 more ones than ten thousands?

(A) 64,786   (B) 64,989   (C) 63,297   (D) 64,898

24 Jaime buys 37 sunglasses. Each pair of sunglasses costs $14. Give an estimate for the total cost of the sunglasses.

(A) $860   (B) $400   (C) $600   (D) $725

25 A box of chocolates contains 72 chocolate pieces. Hattie bought 6 boxes and Erin bought 11 boxes. How many individual chocolate pieces did both Hatti and Erin buy altogether?

(A) 1,129   (B) 1,567   (C) 1,224   (D) 1,112

COMPREHENSIVE ASSESSMENT

1 Comprehensive Assessment

26 Daisy creates the patterns 7, 11, 15, 19, 23, and 27. What are the next five numbers in Daisy's pattern?

- (A) 31, 35, 39, 43, and 47
- (B) 29, 30, 33, 39, and 42
- (A) 26, 31, 34, 37, and 42
- (D) 32, 34, 36, 39, and 44

27 The first number in a sequence is 320. The rule is "multiply by 2, then subtract 200". What number will be third in the sequence?

28 Which fraction is equivalent to $\frac{4}{48}$?

- (A) $\frac{40}{480}$
- (B) $\frac{40}{48}$
- (C) $\frac{40}{400}$
- (D) $\frac{4}{10}$

29 Find the sum $\frac{9}{13} + \frac{7}{13}$.

- (A) $\frac{13}{9}$
- (B) $\frac{16}{13}$
- (C) $\frac{7}{13}$
- (D) $\frac{4}{10}$

30 How much time has passed between 5:38 p.m and 10:42 p.m ?

- (A) 4:55
- (B) 5:32
- (C) 5:04
- (D) 6:55

COMPREHENSIVE ASSESSMENT

Comprehensive Assessment 1

31 One day Elena stitched 4 meters and 20 centimeters of a long cloth. On the next day, she stitched 6 meters and 55 centimeters of this cloth.

_____

_____

32 Find the perimeter of the shape.

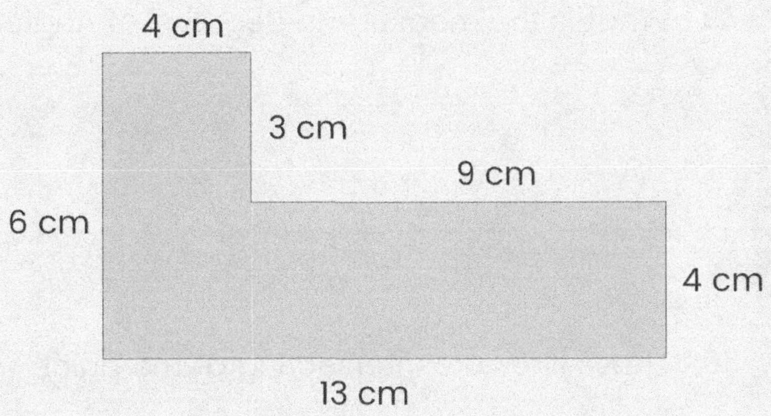

- (A) 39 cm
- (B) 33 cm
- (C) 30 cm
- (D) 35 cm

(4 cm, 3 cm, 9 cm, 6 cm, 4 cm, 13 cm)

33 The measure of angle PQR is 170. Adam states that the measure of Angle SQR is 54. Do you agree with Adam?

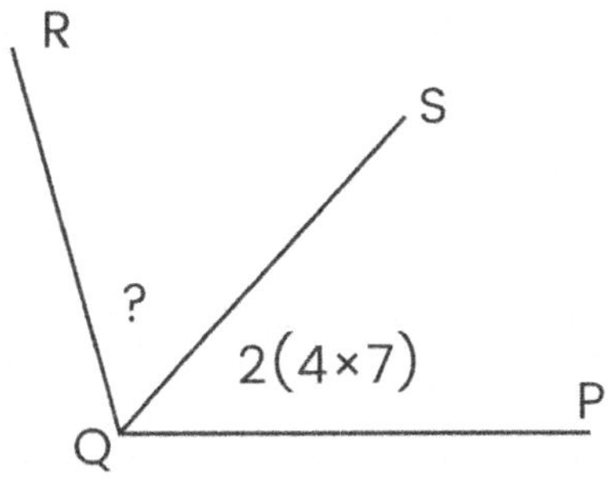

_____

_____

(angle labeled 2(4×7) and ?)

COMPREHENSIVE ASSESSMENT

1 Comprehensive Assessment

34 Draw a shape with these characteristics:
- The shape has 4 sides.
- The length of the 4 sides are equal

35 What is the measure, in degrees, of Angle a?

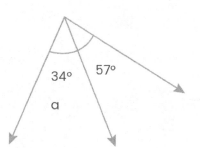

36 These lines are intersecting at the point

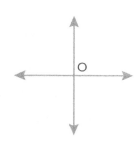

37 What is the measure of Angle E, if the total angles equal 76°?

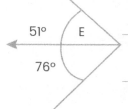

COMPREHENSIVE ASSESSMENT

Comprehensive Assessment 1

38. Find the area of the figure given that each unit on the grid equals 1 cm.

   (A) 79 cm²   (B) 88 cm²
   (C) 80 cm²   (D) 67 cm²

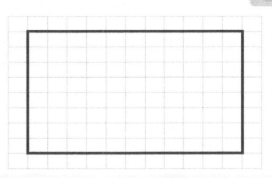

39.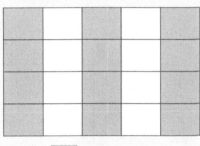

   Write a multiplication sentence that gives the area of the unshaded part of the rectangle.

   _____

   _____

   ☐ 1 Sq. Inch

40. Find the length and perimeter of a rectangle for the given area 255 cm² with the width as 15 cm.

   Length = _____

   Perimeter = _____

41. The perimeter of a square is 129 inches. The length of its side is z inches. Write the equation that can be used to find the value of z.

   _____

   _____

COMPREHENSIVE ASSESSMENT

1 Comprehensive Assessment

42 **True or False:** The dotted line in the image below is a line of symmetry.

Ⓐ True Ⓑ False

43 The graph below shows the percentage of people who chose a different color as their favorite. Which of the following statements is true?

Ⓐ Most people chose green

Ⓑ Most people chose blue

Ⓒ Most people chose black

Ⓓ None of the above is true

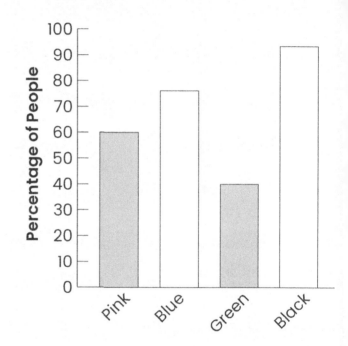

COMPREHENSIVE ASSESSMENT

Comprehensive Assessment 1

44 Use the data in the table to complete the line plot. Plot the frequency of each value above the line.

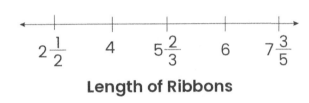

Length of Ribbons					
Ribbon Length (inches)	$2\frac{1}{2}$	4	$5\frac{2}{3}$	6	$7\frac{3}{5}$
Number of ribbons	3	2	4	2	3

Length of Ribbons

45 State True or False: The letter "H" has 3 lines of symmetry.

(A) True (B) False

COMPREHENSIVE ASSESSMENT - 2

COMPREHENSIVE ASSESSMENT

2 Comprehensive Assessment

1 The number 120 is 10 times 12.
Write this as a multiplication equation.

2 There are 7 boxes of ice cream with 24 ice cream bars in each box. 94 children each took an ice cream bar at dinner. How many ice cream bars are left?

Ⓐ 74   Ⓑ 86   Ⓒ 108   Ⓓ 124

3 Zeke collected 325 red balls, 265 blue balls, and 346 green balls. He puts the red balls, blue balls, and green balls together in bags. Each bag has 14 red balls, 12 blue balls, and 10 green balls. How many complete bags did Zeke create?

4 Ben bought cucumbers, carrots, mushrooms, and radishes. This table shows the portion of each vegetable, represented as a fraction. Write an expression to represent the total number of mushrooms and radishes.

Vegetables	Fraction
Cucumber	$\frac{5}{10}$
Carrot	$\frac{15}{100}$
Mushrooms	$\frac{40}{100}$
Radish	$\frac{8}{10}$

COMPREHENSIVE ASSESSMENT

Comprehensive Assessment 2

5 Which is more, 1690 meters or 2 kilometers?

_____

_____

6 A square carpet has sides that are 6 feet long. What is the carpet's area?

(A) 36 sq. ft.   (B) 24 sq. ft.   (C) 18 sq. ft.   (D) 12 sq. ft.

7 Ariel is 10 years old. She is 2 times older than her 5-year-old brother. Write an equation that matches this story. Explain your reasoning.

_____

_____

8 The zoo currently has 25,000 people visiting. The number of people in the amusement park is eight times that of the zoo. How many people are visiting the amusement park?

COMPREHENSIVE ASSESSMENT

2 Comprehensive Assessment

9 The population of Iceland is 6,521,427.
How would this number be expressed in expanded form?

_____

10 Amir travels 657 miles in 9 hours driving at the same speed.
What is his driving speed?

(A) 68 mph (B) 76 mph (C) 73 mph (D) 56 mph

11 A recipe requires $\frac{5}{12}$ cups of wheat flour and $\frac{1}{12}$ cups of butter to make 10 cookies.
Zoe wants to double the recipe. How much flour and sugar will she need?

_____

12 State True or False. The perimeter of the following shape is 16 cm.

(A) True (B) False

5 cm
3 cm   3 cm
5 cm

COMPREHENSIVE ASSESSMENT

Comprehensive Assessment 2

13. Ashley earned $82 today by selling products in her shop. That is 2 times as much as she earned yesterday. How much did Ashley earn yesterday?

_____

_____

14. Dana has 11 boxes of colored pencils. Each box has 10 colored pencils in it. Explain how to express this as a multiplication equation.

_____

_____

15. There are 10 rooms on the first floor. Each room can have up to eight students. What is the maximum number of students that could stay on the first floor?

Ⓐ 70 Ⓑ 80 Ⓒ 90 Ⓓ 100

16. The fifth number in the sequence is 67. The rule is "add 4." What is the first number in the sequence?

_____

_____

COMPREHENSIVE ASSESSMENT

2 Comprehensive Assessment

17 How would the number 7,264,589 be expressed in written form?

_____

_____

18 Which list shows fractions listed from greatest to least?

Ⓐ $\frac{8}{9}, \frac{5}{7}, \frac{1}{4}$ Ⓑ $\frac{1}{4}, \frac{5}{7}, \frac{8}{9}$ Ⓒ $\frac{5}{7}, \frac{1}{4}, \frac{8}{9}$ Ⓓ $\frac{1}{4}, \frac{8}{9}, \frac{5}{7}$

19 Jenny makes $9 an hour. How much money does she make in a whole month, if she works 4 hours every day of the 30-day month?

Ⓐ $1,560 Ⓑ $1,340 Ⓒ $1,080 Ⓓ $1,180

20 A chocolate cake costs $13. How many chocolate cakes can be bought with $208?

Ⓐ 12 Ⓑ 16 Ⓒ 18 Ⓓ 20

21 Hope has 65 water bottles. Jada has 4 times as many water bottles as Hope. How many water bottles does Jada have?

Ⓐ 340 Ⓑ 380 Ⓒ 220 Ⓓ 260

COMPREHENSIVE ASSESSMENT

Comprehensive Assessment 2

22 Twenty-two watermelons cost $198.
Use r for the cost of one watermelon.

(A) r = $9   (B) r = $8   (C) r = $11   (D) r = $7

23 Douglas is thinking about a number with the following characteristics:
- 1 ten thousands
- 8 thousands
- 4 hundreds
- 2 tens
- 6 ones

What number is Douglas thinking about?

(A) 18,546   (B) 18,426   (C) 19,245   (D) 19,317

24 Nick sold 199 pies. Each pie costs $8. Which of the following is closest to the total amount that Nick received?

(A) $5,000   (B) $1,000   (C) $2,000   (D) $7,000

25 Tanya's shop has 5,000 mushrooms. She will pack 25 mushrooms in each box. She will sell each box for $45. Determine the amount of money Tanya could make if she sells all boxes of mushrooms.

(A) $9,000   (B) $10,000   (C) $8,000   (D) $11,000

COMPREHENSIVE ASSESSMENT

2 Comprehensive Assessment

26 The first two terms of a pattern are 3 and 8. A third term is obtained by adding the previous two terms. Find the first five terms of this pattern.

(A) 3, 8, 18, 29, and 35

(B) 8, 3, 11, 9, and 25

(C) 3, 8, 12, 19, and 32

(D) 3, 8, 11, 19, and 30

27 What rule is used to create this pattern? Explain how you know.
5, 10, 20, 40, 80.

_____

_____

28 The shaded part of the model below represents a fraction of the total area of the model.

Which of the following sets of fractions are equivalent to the shaded portion in the model?

(A) $\frac{2}{4}, \frac{12}{20}, \frac{24}{30}$

(B) $\frac{1}{3}, \frac{16}{21}, \frac{24}{36}$

(C) $\frac{1}{2}, \frac{12}{24}, \frac{18}{36}$

(D) $\frac{2}{3}, \frac{18}{45}, \frac{24}{65}$

COMPREHENSIVE ASSESSMENT

Comprehensive Assessment 2

29 Find the difference $\frac{14}{17} - \frac{10}{17}$

(A) $\frac{17}{4}$ (B) $\frac{4}{17}$ (C) $\frac{3}{17}$ (D) $\frac{17}{6}$

30 Eugene spends 90 minutes walking every day. How many hours does he spend walking each week if he walks every day of the week?

(A) 12.2 hours   (B) 12.67 hours
(C) 11.68 hours   (D) 10.5 hours

31 The height of the main door is 9 meters. What is the height of the door in millimeters?

(A) 9,000 mm   (B) 5,000 mm   (C) 6,000 mm   (D) 8,000 mm

32 The measure of angle G is 120°. What fraction of a complete circle is the measure of angle G?

_____

_____

COMPREHENSIVE ASSESSMENT

2 Comprehensive Assessment

33 This line plot shows the number of people per cup of orange juice, 10 people drink in the shop. How many people drink less than 2 cups of juice?

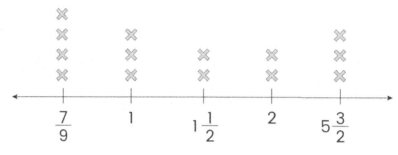

Number of People Per Cup Orange Juice

34 Draw a shape or figure with exactly 4 lines of symmetry.

35 The measure of Angle GDF is 23 degrees. The measure of Angle FDE is four times as large as Angle GDF. What is the measure, in degrees, of Angle GDE?

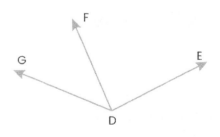

COMPREHENSIVE ASSESSMENT

Comprehensive Assessment 2

36 What is the area in units of the shaded shape?t

_____

_____

(Figure with dimensions: 7, 2, 13, 27)

37 Determine the figure's area given that each unit on the grid is equal to 1 cm in length.

- A) 124 cm²
- B) 172 cm²
- C) 156 cm²
- D) 184 cm²

38 Mary wants to build a house in a rectangular area of 48 square yards. What is the width of the house if the length is 6 yards?

- A) 4 yards
- B) 6 yards
- C) 10 yards
- D) 8 yards

39 The perimeter of the square-shaped cloth is 60 inches. Alex decided to extend the cloth by doubling the sides. What is the new area?

- A) 900 square inches
- B) 700 square inches
- C) 600 square inches
- D) 1,000 square inches

COMPREHENSIVE ASSESSMENT

2 Comprehensive Assessment

40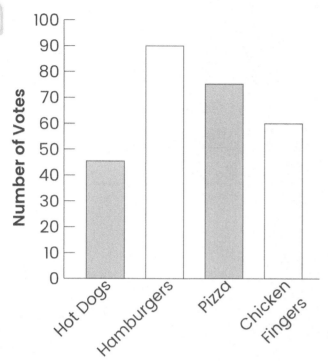

A survey was conducted to find out the favorite food among a group of students.

How many students said their favorite food was a hamburger?

A) 60 B) 45
C) 90 D) 75

41 The school is 100 m long and has an area of 1800 m². The school hostel has the same area but is two times longer. What is the perimeter of the school hostel?

A) 418 m   B) 478 m   C) 512 m   D) 564 m

42 True or False: The dotted line in the image below is a line of symmetry.

A) True   B) False

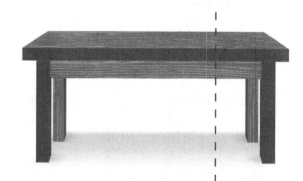

COMPREHENSIVE ASSESSMENT

Comprehensive Assessment 2

43. Which of the following capital letters has at least one line of symmetry?

    (A) F   (B) N   (C) O   (D) S

44. How many pairs of parallel lines exist in this hexagon?

    _____

    _____

    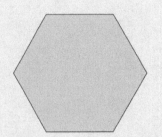

45. Which shape is a rhombus?

    (A)   (B)   (C)   (D)

ANSWERS AND EXPLANATIONS

ANSWERS AND EXPLANATIONS

1. MULTIPLICATION AND DIVISION

1.1 Multiplication Properties and Facts

1. Answer: D
Explanation: Solve this problem with multiplication:
21 times 15 is 315.
(21 × 15 = 315)

2. Answer: $3,675
Explanation: Multiplying half the student population (1,225) by the cost of each lunch ($ 3) is $ 3,675.

3. Answer: A
Explanation: The minimum total points would occur if the team scored the minimum number of points (23) in every game. Since there are 18 games in a season, the minimum total score would be 23 × 18 = 414 points. The maximum total points would occur if the team scored the maximum number of points (38) in every game. Therefore, the maximum total score would be 38 × 18 = 684 points.

4. Answer: $ 2,000
Explanation: One-third of 120 is 40. There are 40 adults and 80 children in the group. Using the expression (40×20) +(80×15). The total amount spent is $2,000.

5. Answer: C
Explanation: Using a rounding and decomposition strategy, 32×15 can be calculated as (30×15) + (2×15)

6. Answer: B
Explanation: 5×7 = 35
7×5 = 35
5×7 = 7×5

7. Answer: A
Explanation: Each tray makes 12 glasses of apple juice.
5 trays = 5×12
Using the distributive property 5×10+5×2
= 50+10 = 60

8. Answer: B
Explanation: One hour = $ 4
Per day 5 hours = 5×$4 = $20
One week = 7×20 = 140

9. Answer: C
Explanation: One blue chair costs $ 7.
One green chair costs $7.
Noah bought 5 blue chairs = 5×7 = 35
Noah bought 4 green chairs = 4×7 = 28
35 + 28 = 63

10. Answer: D
Explanation: 1 crate of mangoes containing 13 mangoes.
4 crates of mangoes containing = 4×13 = 52
1 crate of apples, containing 13 apples.
8 crates of apples containing = 8×13=104
52+104 = 156

11. Answer: B
Explanation: 1 box has 12 pens.
9 boxes = 9×12=108

ANSWERS AND EXPLANATIONS

12. Answer: D
Explanation: A rectangle has 4 sides.
11 rectangles have=4×11=44

13. Answer: C
Explanation: One cage has 3 dogs.
5 cages have 5×3=15

14. Answer: A
Explanation: One box has 18 oranges.
9 boxes have 9×18=162

15. Answer: B
Explanation: A shopkeeper packs 8 sandwiches in a box.
12 boxes have =8×12=96

1.2 Understanding Division
---

1. Answer: D
Explanation: The solution is determined by decomposing 752 into 2 values that are divisible by 8: 752 can be decomposed into 752+ 8.
Then, divide each value by 8.

2. Answer: C
Explanation: Divide 426 by 14 which gives 30 with a remainder of 6.
William fills 30 baskets with 14 bananas each, and one basket with 6 bananas. So, he needs 31 baskets.

3. Answer: 31
Explanation: The number 415 divided by 13 is 31, with a remainder of 12.
This means there will be 31 complete teams.

4. Answer: 172 tickets
Explanation: Divide the amount of money collected by the price of each ticket. The high school sold 172 tickets.
(1,548 ÷ 9 = 172)

5. Answer: B
Explanation: There are enough books, notepads, and pencils to create 42 complete bags.
The limiting item is the books.

6. Answer: 146
Explanation: The team sells 1,596 jerseys at a cost of $8 each.
Multiplying these two values gives $12,768.
Subtraction of $12,768 from total earnings ($14,520) yields the amount earned from selling ball racks ($1,752).
The number of ball racks sold is calculated by dividing $1,752 by 12.
The team sold 146 ball racks.

7. Answer: D
Explanation: To find the quotient of 16 and 4, one must look for a number that, when multiplied by 4 gives 16.

8. Answer: A
Explanation: At a picnic, 104 people want to eat lunch.
Each picnic table seats 13 people.
The number of tables that the people need=104÷13=8

ANSWERS AND EXPLANATIONS

9. Answer: C
Explanation: Logan caught 81 lobsters. Each hour, he caught nine lobsters. Number of hours he spent fishing = 81÷9 = 9

10. Answer: B
Explanation: One dragon fruit costs $6. Number of dragon fruits bought for costs $72 =72÷6=12.

11. Answer: A
Explanation: A basket contains 105 carrots. Amelia wants to equally share a basket of 105 carrots with 15 students. Each student gets =105÷15=7 carrots.

12. Answer: Another way to arrange the 64 bicycles is to have 8 rows with 8 bicycles in each row.
Explanation: 64 can be expressed as 8×8. Another way to arrange the 64 bicycles is to have 8 rows with 8 bicycles in each row.

13. Answer: C
Explanation: Each player receives 3 towels. So the total number of towels is =27×3=81. Number of boxes =81÷9=9 boxes.

14. Answer: D
Explanation: The total number of pens=6×11=66.
Dylan arranged his pen into 11 groups. The number of pens in each group =66÷11=6 pens.

15. Answer: B
Explanation: Hannah walks 4 km a day. Number of days she has walked a total of 32 km=32÷4=8.

1.3 Comparison of Multiplication and Division

1. Answer: C
Explanation: Kevin is 7 years old. Colton is 10 times as old as Kevin. Colton's age =7×10=70

2. Answer: B
Explanation: Robert paid for notebooks =$ 15. Robert paid 5 times as much for school fees=15×5=75

3. Answer: C
Explanation: Parker paid for a uniform=$26. A uniform is 2 times as expensive as a pair of shoes =26÷2 The cost of a pair of shoes =$13

4. Answer: B
Explanation: Last month Adam saved =$7
This month Adam saved three times as much as last month =7×3=$21

5. Answer: A
Explanation: Number of dogs in pet shop =66. Number of cats in pet shop =11 Number of times more dogs are there than cats =>66÷11=6

ANSWERS AND EXPLANATIONS

6. Answer: C
Explanation: A monkey weighs =72 kg.
A rabbit weighs = 9 kg.
Number of times a monkey is as heavy as a rabbit => 72÷9=8

7. Answer: D
Explanation: A television costs =$250.
A smartphone costs 3 times as much as a television =250×3=750
A smartphone costs =$750

8. Answer: A cow weighs 14 times as heavy as a rabbit.
Explanation: A cow weighs =140 kg.
A rabbit weighs=10 kg.
A cow weighs 14 times as much as a rabbit.

9. Answer: C
Explanation: 50 balls are 5 times as many as 10 balls.
5×10=50

10. Answer: A
Explanation:
The price of a medium-size cake=$8
The price of a jumbo-size cake=$32
Number of times a medium cake is cheaper than a jumbo cake => 32÷8=4

11. Answer: C
Explanation: 1 kg of mango=$36
1 kg of banana=$6
Number of times 1 kg of mango is more expensive than 1 kg of Banana=36÷6=6

12. Answer: B
Explanation: Jasmine's score on an English exam =36 points
Naomi's score is 2 times as many points as Jasmine's score =36×2=72
Naomi's score on an English exam =72 points.

13. Answer: A
Explanation: Mary has 40 red roses.
Lilly has 6 times as many red roses as Mary =40×6=240.
Lilly has 240 red roses.

14. Answer: D
Explanation: The length of white cloth = 90 meters
White cloth is 6 times longer than a pink cloth = 90÷6=15
The length of the pink cloth = 15 meters

15. Answer: B
Explanation: Cora sold 28 pens.
Cora sold 7 times as many pens as Ivan =28÷7=4 Ivan sold 4 pens.

1.4 Equations for Multiplication and Division

1. Answer: A
Explanation: x stands for a number.
$x \times 7 = 63$
$7 \times x = 63$
$63 \div 7 = x$

ANSWERS AND EXPLANATIONS

2. Answer: B
Explanation: The quotient of 54 and 9 is q. 54÷9=q. q=6

3. Answer: D
Explanation: Alexa drives 20 miles to the beach.
Bella travels x miles to get to the beach.
The given equation 20÷x=4
20÷4=x, x=5
5 miles. Alexa drives 4 times as many miles as Bella.

4. Answer: B
Explanation: Sarah sold 25 cups of green tea at her shop.
Riley's shop sold n cups of green tea.
The equation 25−10=n
n=15
Riley sold 10 fewer cups than Sarah.

5. Answer: A
Explanation: A bakery has an order for 85 packs of bread.
They bake 5 packs of bread per hour.
Number of hours a bakery needs to finish the order => 85÷5=p
p=17

6. Answer: C
Explanation: Lucy bought 4 boxes of cakes.
Each box contains 18 cakes.
Number of cakes Lucy bought => 4×18=y
y=72

7. Answer: B
Explanation: A number r is multiplied by 3 to get the result 21.
r×3=21
r=7

8. Answer: A
Explanation: There are 6 triangles in each row.
Number of triangles in three rows =6×3=18

9. Answer: B.
Explanation: The given expression is (2×z)×6.
2×2×6=24
z=2

10. Answer: C
Explanation: Seven people equally divided the money earned by selling baseball bats. Each person received y dollars.
The total amount of money the seven people earned is $98. Money received by each of them => 7×y=98
y=14

11. Answer: A
Explanation: 5×6=30
48÷6=8
2×6=12

ANSWERS AND EXPLANATIONS

12. Answer: C
Explanation: Let r be the total number of pencils. Number of boxes of pencils Rachel bought =11
Each box contains 12 pencils.
Total number of pencils in 11 boxes
=>11×12=132

13. Answer: B
Explanation: Let x be the number of months.
Victor is saving money to buy a new tablet =$600
He plans to save $60 each months.
To find number of months we have to divide total amount by amount per month.
600÷60=x
x=10

14. Answer: D
Explanation: Let p represents the cost of each tree.
Alan bought and planted 9 trees in his garden.
He paid $45 for all nine trees.
The cost of one tree =45÷9=p
p=5.

15. Answer: C
Explanation: Let q be the cost of one book.
Fifteen books =$225
The cost of one book =15×q=225
q=15

1.5 Chapter Review

1. Answer: 432 cookies.
Explanation: Mya ordered 12 packets of cookies.
Each packet contains 36 cookies.
Total number of cookies in 12 packets
=12×36=432.

2. Answer: A
Explanation: A theater has 4 sections.
Same number of seats in each section.
Audience in the morning show =1,752
Number of audience in each section
=1,752÷4=438

3. Answer: Mr. Bryan's family spends more money on a trip.
Explanation: Number of people in Mr. Bryan's family=21
Number of people in Mr. Steven's family =16. Each person in Mr. Bryan's family pay for a trip to the aquarium=$25
Each person in Mr. Steven's family pay for a trip to the national park=$32
Mr. Bryan's family spends=21×$25=$525
Mr. Steven's family spends=16×$32=$512
Mr. Bryan's family spends more money on a trip.

4. Answer: D
Explanation: Total number of lines = 24+18=42
Time for typing each line = 1 minute
Total time taken = 42×1=42 minutes

ANSWERS AND EXPLANATIONS

5. Answer: B
Explanation: Number of cupcakes Mark sold in a day=24
Mateo sold 3 times as many cupcakes as Mark.
Number of cupcakes Mateo sold in the same day=24×3=72.

6. Answer: C
Explanation: Number of students in a school bus=40
Total number of students=8000
Number of buses needed =8000÷40=200

7. Answer: A
Explanation: Number of sugar cane plants in the garden=120,000
Each day 500 sugar cane plants are cut to make sugar cane juice.
Number of days to cut all sugar cane plants are = 120,000÷500 = 240 days.

8. Answer: B
Explanation: Total number of chocolates = 50
Number of children = 10
The operation that gives the correct number of shares = 50÷10

9. Answer: A
Explanation: Number of boxes of water bottles Emma bought =4
Each box contains 20 water bottles.
Total number of water bottles =20×4=80

10. Answer: D
Explanation: One bookshelf has 30 books.
16 bookshelves have =30×16=480

11. Answer: 9 complete teams.
Explanation: The number 139 divided by 15 is 9, with a remainder of 4.
This means there will be 9 complete teams

12. Answer: C
Explanation: Jacob drinks 3 litres of water per day.
Number of days he needs to drink 69 litres of water =69÷3=23

13. Answer: B
Explanation: Number of apple trees in the garden=120
Number of orange trees in the garden =60
Number of times more apple trees than orange trees =>120÷60=2

14. Answer: A
Explanation: Number of exercises Rose solved in the math book=7
Jasmine solved 4 times as many exercises as Rose solved =7×4=28
Number of exercises Jasmine solved =28

15. Answer: C
Explanation: n stands for a number.
n×11=121
11×n=121
121÷11=n

ANSWERS AND EXPLANATIONS

16. Answer: D
Explanation: A hexagon has 6 sides.
14 hexagons have =6×14=84

17. Answer: 8,192
Explanation: A building has 128 floors.
Each floor has 64 apartments.
Total number of apartments = 128×64 = 8,192.

18. Answer: C
Explanation: Alice caught 36 fish.
Mya caught x number of fish.
The given equation is 36×3=x
x=108
Mya caught 3 times as many fish as Alice.

19. Answer: B
Explanation: Amy filled 7 boxes with pineapple.
Each box contains 12 pineapples.
Number of pineapples Amy used
=>12×7=p
p=84

20. Answer: 579÷3
Explanation: Divide the total amount of money $579 to buy light bulbs by the cost of one light bulb.
Mr. Antonio bought 193 light bulbs.

2. PLACE VALUE

2.1 Expanded Form and Numerals

1. Answer: B
Explanation: Since there are no candies in the thousands place (0), adding one thousand candies would result in 1,000 candies being in that place. Therefore, the digit in the thousands place would change from 0 to 1.

2. Answer: A
Explanation: 54,786
Fifty-four thousand seven hundred and eighty-six.

3. Answer: B
Explanation: 89,524
Eighty-nine thousand five hundred and twenty-four.

4. Answer: D
Explanation: Vivian writes a number in her notebook = 69,317

5. Answer: B
Explanation: Total number of fish in an aquarium = 154,387
If one thousand additional fish are added, the digit change in 154,387 is 4.

6. Answer: A
Explanation: 20,000+4,000+600+80+0
The place value of the underlined digit in 24,680 is 1,000s, while the value of that digit is 4,000.

ANSWERS AND EXPLANATIONS

7. Answer: C
Explanation: 876,521
The place value of the digit 5 in 876,521 is hundreds.

8. Answer: D
Explanation: 13,569
The place value of the digit 9 in 13,569 is ones.

9. Answer: B
Explanation: Option B (7<u>7</u>,987) has a 7 in place of ten thousand (70,000).

10. Answer: C
Explanation: Twenty-six thousand nine hundred and sixteen

11. Answer: A
Explanation: John saved a number in his smartphone = 10,953

12. Answer: B
Explanation: 31,387
Thirty-one thousand, three hundred, and eighty-seven.

13. Answer: D
Explanation: 9<u>7</u>6,246
70,000

14. Answer: B
Explanation: 192,713
The number 9 in the ten thousands place and the number 1 in the tens place.

15. Answer: D
Explanation:
400,000+20,000+1,000+900+80+3
The place value of the underlined digit in 421,983 is ones, while the value of that digit is 3.

2.2 Friendly And Benchmark Numbers

1. Answer: A
Explanation: Rounding to the nearest 1,000.
4,000+3,000=7000

2. Answer: C
Explanation: Hint: 0.39 is close to 25
Rounding 242 to the nearest 100 gives 200.
$(2 \div 5) \times 200 = 80$

3. Answer: B
Explanation: Rounding 527 to the nearest 100 gives 500.
Rounding 478 to the nearest 100 gives 500.
527 and 478 give the same result when rounded to the nearest 100.

4. Answer: D
Explanation: 54 is the largest whole number that gives 50 when rounded the nearest nearest 10

5. Answer: A
Explanation: 66 is the smallest whole number that gives 70 when rounded to the nearest 10.

ANSWERS AND EXPLANATIONS

6. Answer: A
Explanation: True

7. Answer: C
Explanation: Daniel buys 31 watches.
Rounding the numbers 31 and 16 to the nearest 10.
30×20=$600

8. Answer: B
Explanation: Rounding 47 to the nearest 10 is 50.
Rounding 741 to the nearest 100 is 700.
50×700=35,000

9. Answer: C
Explanation: Rounding 526 to the nearest 100 is 500.
Rounding 22 to the nearest 10 is 20.
500÷20=25

10. Answer: A
Explanation: Rounding $93,658 to the nearest 1,000 gives $94,000

11. Answer: D
Explanation: Estimate:
$16+$11+$22+$19+$27
Rounding each number to the nearest 10.
$20+$10+$20+$20+$30=$100

12. Answer: B
Explanation: Sofia sold 398 balls.
Each ball costs $7.
398×7=$2,786
$3,000 is the closest amount.

13. Answer: B
Explanation: To round 425 to the nearest 500, we need to determine which multiple of 500 it is closest to.
The midpoint between 400 and 500 is 450. Since 425 is closer to 400 than it is to 500, it would round down to 400, not 500.

14. Answer: D
Explanation: Rounding 62,756 to the nearest 1,000 gives 63,000.
Rounding 541 to the nearest 100 gives 500
63,000÷500=126

15. Answer: A
Explanation: Rounding 325,654 to the nearest 100,000 is 300,000
Rounding 23,851 to the nearest 10,000 is 20,000
300,000-20,000=280,000

2.3 Multiply and Divide by Multiples of 10
------

1. Answer: C
Explanation: Number of hours Jackson works per day=4
Number of days Jackson works in a month=25
Total number of hours Jackson works in a month =4×25=100 hours

ANSWERS AND EXPLANATIONS

2. Answer: B
Explanation: A box contains 35 balls.
Number of boxes Harris bought = 7
Number of boxes Ian bought = 9
Number of balls Harris bought = 7×35 = 245
Number of balls Ian bought = 9×35 = 315
Number of balls both Harris and Ian bought = 245+315 = 560

3. Answer: D
Explanation: Number of rooms in a hostel = 250
Each room has 6 students.
Total number of students in a hostel = 6×250 = 1,500

4. Answer: A
Explanation: Every month Max saves = $50
He saves the amount for 10 months = $50×10 = $500

5. Answer: B
Explanation: Number of chocolates in the bag = 640
Number of donuts in the bag = 640÷8 = 80

6. Answer: D
Explanation: 7,540÷13 = 580
The number x is 580.

7. Answer: B
Explanation: Number of flowers Jack bought = 500
Cost of one rose = $6
Cost of one lily = $10
$6+$10 = $16
16×500 = 8000

8. Answer: C
Explanation: Cost of one ring = $60
Number of rings Ivan bought = 4,200÷60 = 70

9. Answer: A
Explanation: 14×10 = 140

10. Answer: D
Explanation: 80×10 = (8×1)×100 = 800

11. Answer: B
Explanation: For 1 hour to prepare food = 5,000÷5 = 1,000
Number of hours it takes to prepare food for 15,000 people = 15,000÷1,000 = 15 hours

12. Answer: C
Explanation: 40×7 = 280
4×10×7 = 280

13. Answer: A
Explanation: Jack packs 100 cakes per hour.
100 = 1 hours = 60 min
Minutes take to complete 350 cakes = 60×350÷100 = 21,000÷100 = 210 min

ANSWERS AND EXPLANATIONS

14. Answer: B
Explanation: Lucy's shop has 3,000 pears. Each box contains 30 pears. Each box costs $20.
3,000÷30=100
100×$20=$2,000

15. Answer: D
Explanation: Number of school buses at the school =50
Each bus contains 20 seats.
Number of seats in 50 buses
=50×20=1,000

2.4 Chapter Review

1. Answer: D
Explanation: The value 5 is in thousands place.

2. Answer: C
Explanation: Fifty-six thousand, two hundred, and eighty-four.

3. Answer: A
Explanation: The value 165,825 is expanded in
100,000+60,000+5,000+800+20+5

4. Answer: B
Explanation: 22,020

5. Answer: 7 hundred, 2 tens and 0 ones
Explanation: ten thousand seven hundred and twenty =1 ten thousand+7 hundred+2 tens+0 ones

6. Answer: C
Explanation: The number smallest to largest are 622, 699, 701, and 754

7. Answer: D
Explanation: Numbers in descending order: 1,100, 987, 875, 798, 748, 652, 610, 547, 387, 88, 12

8. Answer: B
Explanation: Sam wrote a number 71,584.

9. Answer: A
Explanation: 97,359

10. Answer: C
Explanation: Rounding 920 to the nearest 100 is 900.

11. Answer: B
Explanation:
7×10,000+5×1,000+3×100+1×10+9=75,319

12. Answer: D
Explanation: The underlined digit worth $1,000 is 1,578.

13. Answer: B
Explanation: Rounding 26,780 to the nearest 1,000 gives 27,000. Rounding 195 to the nearest 100 gives 200.
27,000÷200=135

14. Answer: C
Explanation: Rounding 6,348 to the nearest 1,000 gives 6,000. Rounding 5,956 to the nearest 1,000 gives 6,000

ANSWERS AND EXPLANATIONS

15. Answer: A
Explanation: Number of crabs Susan cooked=48
Cost of one crab =$6
Total costs of crab =48×$6=$288

16. Answer: B
Explanation: Estimate:
$24+$22+$28+$20
Rounding each number involved to the nearest 10.
$20+$20+$30+$20=$90

17. Answer: D
Explanation: 98,560÷140=704
The number x is 704

18. Answer: C
Explanation: A candle machine produces 500 candles per hour.
500 candles=1 hours=60 min
Minutes it takes to produce 5,500 candles = 60×5,500÷500 = 330,000÷500 = 660 min

19. Answer: A
Explanation: The number 6 is the thousands place.

20. Answer: B
Explanation:
800,000+10,000+3,000+500+60+2= 813,562

3. FACTORS AND PATTERNS

3.1 Relationship Between Factors and Multiples

1. Answer: A
Explanation: The number 36 has 4 factor pairs: 2 &18, 3 &12, 4 & 9, and 6 & 6.

2. Answer: C
Explanation: The number 52 cannot be divided equally by 8.
Dividing 52 by 8 leaves a remainder of 4.

3. Answer: D
Explanation: The factors of 81 are 1, 3, 9, 27, and 81.

4. Answer: B
Explanation: Dividing 162 by 9 results in 18 with no remainder.
Therefore, 9 is a factor of 162.

5. Answer: D
Explanation: The number 102 divided by 4 results in 20 and a remainder of 2. Thus, 4 is not a factor of 102.

6. Answer: 1&76, 2&28, and 4&19
Explanation: Answers must include a reasonable explanation such as, "The factor pairs for 76 are 1&76, 2&28, and 4&19.

7. Answer: 14, 28, 42, 56, and 70
Explanation: The first five multiples of 14 are 14, 28, 42, 56, and 70.

ANSWERS AND EXPLANATIONS

8. Answer: D
Explanation: 4 is a number that is a multiple of 16 and a factor of this number.

9. Answer: C
Explanation: Number 11 is a factor of 33. But number 11 is not a multiple of 3

10. Answer: B
Explanation: $2 \times 15 = 30$ and $30 \times 2 = 60$

11. Answer: A
Explanation: The numbers 18, 27, 45, and 72 are divisors of 9 and also multiples of 3.

12. Answer: A
Explanation: $4 \times 15 = 60$. 60 is a multiple of 4 and 15, 15 and 4 are divisors of the number 60.

13. Answer: 7, 6
Explanation: If he arranges the color pens in 6 rows, there will be 7 color pens in each row.
If he arranges the colored pens in 6 rows, there will be 7 colored pens in each row.

14. Answer: 7, 21, and 21, 7
Explanation: 7 is a factor of 21
21 is a multiple of 7

15. Answer: Prime
Explanation: Prime numbers have exactly 2 factors, 1 and the number itself.

3.2 Predict and Extend Growing and Repeating Patterns

1. Answer: A
Explanation: $2+7=9$
$9+7=16$
$16+7=23$
$23+7=30$

2. Answer: C
Explanation: $17 \times 3 = 51$
$51 \times 3 = 153$
$153 \times 3 = 459$
$459 \times 3 = 1,377$
$1,377 \times 3 = 4,131$

3. Answer: 14, 60, 244, 980, and 3,924
Explanation: a. 14
b. $14 \times 4 + 4 = 60$
c. $60 \times 4 + 4 = 244$
d. $244 \times 4 + 4 = 980$
e. $980 \times 4 + 4 = 3,924$

4. Answer: A
Explanation: $(22+5) \times 2 = 54$

5. Answer: B
Explanation: Multiply 5 and subtract 2
$2 \times 5 - 2 = 8$
$4 \times 5 - 2 = 18$
$6 \times 5 - 2 = 28$
$8 \times 5 - 2 = 38$

6. Answer: D
Explanation: $23+5=28$
$28+3=31$
$31+5=36$

ANSWERS AND EXPLANATIONS

36+3=39
39+5=44

7. Answer: B
Explanation:
3+6=9
9+6=15
15+6=21
21+6=27

8. Answer: A
Explanation: The first two terms are 1 and 3. Third term =1×3=3. Fourth term =3×3=9,. Fifth term =3×9=27. Sixth term =9×27=243

9. Answer: C
Explanation: At first he made 352
352+19=371. 371+19=390. 390+19=409
409+19=428

10. Answer: B
Explanation: On Sunday 140 visitors
Monday =140+25=165
Tuesday =165+25=190
Wednesday =190+25=215
Thursday =215+25=240

11. Answer: D
Explanation: 900×5=4,500
4,500×6=27,000

12. Answer: A
Explanation: 7+17=24

13. Answer: C
Explanation: 2×4=8

14. Answer: B
Explanation: Day 1 =125
Day 2 =125+15=140
Day 3 =140+15=155
Day 4 =155+15=170

15. Answer: A
Explanation: Day 1 =12
Day 2=12+11=23
Day 3 =23+11=34
Day 4 =34+11=45

3.3 Pattern Rules

1. Answer: C
Explanation: Multiply the number of hours by 70.
For 13 hours, the number of kilometers Rose traveled =13×70=910

2. Answer: B
Explanation: The rule of the pattern is to multiply the last number by 4 to find the next number.

3. Answer: C
Explanation: The one digit will be 0, 1, 2, 3, and so on.
This means that the numbers will alternate between even and odd.

4. Answer: A
Explanation: The one digit will be 0, 2, 4, 6, and 8.
This means that the numbers will be even.

ANSWERS AND EXPLANATIONS

5. Answer: B
Explanation: The rule of the pattern is to add 8. The next number is 40.

6. Answer: D
Explanation: The rule is: add 6. The missing number is 57 because 51+6 = 57.

7. Answer: 93
Explanation: The rule is to subtract 5. The missing number is 93 because 98−5=93

8. Answer: Add 16
Explanation: Using the pattern, 103 − 87 = 16, 87 − 71 = 16, So the rule of the sequence is to add 16.

9. Answer: 76
Explanation: Starting with 28 and adding 12, the sequence is 28, 40, 52, 64, 76, ... The fifth number is 76.

10. Answer: 84
Explanation: The rule is to add 9 (or subtract 9 if moving from right to left) so the number before 93 is 84.

11. Answer: 90
Explanation: The rule is to divide by 3 because 30 / 3 = 10. Therefore, 270 to 90.

12. Answer: 65
Explanation: The sequence starts with 200, so 200÷2=100, and 100+10=110 Then, 110÷2=55, and 55+10=65

13. Answer: 62
Explanation: If the rule of the sequence is to add 15, then, the rule going to the left of the given number is to subtract 15. Beginning with the 4th term of 107; 107 − 15=92, 92 − 15 = 77, and 77−15=62. The first term of the sequence is 62, the second term is 77, the third term is 92, and the fourth term is 107 as given.

14. Answer: Even
Explanation: Adding an even number to other even numbers will result in even numbers.

15. Answer: Add 6
Explanation: The pattern is to add 6 because each number increases by 6

> **3.4 Prime and Composite Numbers**

1. Answer: A
Explanation: The number 53 is not a composite number. It is a prime number because it cannot be divided by any number except 1 and itself.

2. Answer: D
Explanation: The number 108 is not prime because it can be divided by 2, 3, 4, 6, 9, 12, 18, 27, and 36.

ANSWERS AND EXPLANATIONS

3. Answer: C
Explanation: The number 37 is the next prime number after 31 because it cannot be divided by any number except 1 and itself.

4. Answer: Prime
Explanation: Answers must include a reasonable explanation such as, "The number 73 is a
prime number because its only factors are 1 and 73."

5. Answer: 43
Explanation: After 41 is 42, then 43. All even numbers greater than 2 are composite numbers. The number 42 can be divided by 2, and 43 can only be divided evenly by 1 and 43 so it is prime."

6. Answer: Composite
Explanation: All composite numbers have more than one-factor pair.

7. Answer: 5
Explanation: The prime numbers between 11 and 30 are 13, 17, 19, 23, and 29.

8. Answer: 72
Explanation: The number 71 is prime, so 72 is the next composite number after 70.

9. Answer: 10
Explanation: The prime numbers between 101 and 150 are 103, 107, 109, 113, 127, 131, 137, 139, 143, and 149.

10. Answer: Prime
Explanation: The number 101 is prime because its only factor pair is 1 & 101.

11. Answer: B
Explanation: The number 100 is a composite number.

12. Answer: C
Explanation: The first number is 3.
$3 \times 2 + 1 = 7$
$7 \times 2 + 1 = 15$
$15 \times 2 + 1 = 31$
$31 \times 2 + 1 = 63$

13. Answer: Subtract 4
Explanation: The first number is 127
$127 - 4 = 123$
$123 - 4 = 119$
$119 - 4 = 115$
$115 - 4 = 111$

14. Answer: Output = Input ÷2+5
Explanation: $2 \div 2 + 5 = 6$
$4 \div 2 + 5 = 7$
$6 \div 2 + 5 = 8$
$8 \div 2 + 5 = 9$
$10 \div 2 + 5 = 10$

15. Answer: A
Explanation: $36 \div 3 + 10 = 22$
$36 \div 6 + 10 = 16$
$36 \div 9 + 10 = 14$
$36 \div 12 + 10 = 13$
36 hours divided by the number of exercises, and add 10 hours.

ANSWERS AND EXPLANATIONS

3.5 Chapter Review

1. Answer: C
Explanation: The number 120 has 8 factor pairs (1 & 120, 2 & 60, 3 & 40, 4 & 30, 5 & 24, 6 & 20, 8 & 15, 10 & 12).

2. Answer: B
Explanation: Multiplying 5 by 13 equals 65.

3. Answer: Composite
Explanation: The factors of 98 are 1, 2, 7, 14, 49, and 98. Thus, 98 is not prime.

4. Answer: A
Explanation: The pattern is "add 35," so the next number is 286 + 21 = 307.

5. Answer: D
Explanation: The pattern is "divide 5", so the next number is 180 5 = 36.

6. Answer: B
Explanation: The pattern is 19×2=38, 38×2=76.
The rule of the pattern is "multiply by 2."

7. Answer: C
Explanation: The pattern is 873−27=846, 846−27=819, 819−27=792, and 792−27=765
The number that comes before 792 is 765.

8. Answer: 11
Explanation: The order of operations says to divide 81 by 3 first, giving 27. Then, subtract 27 from 16: 27− 16= 11.

9. Answer: 15
Explanation: Add 68 plus 52 and get 120. Then, divide 120 by 8, which equals 15 with a remainder of 0.
Daniel can make 15 full boxes of 8 pens in each box.

10. Answer: C
Explanation: The rule is 5×2+15=25, 10×2+15=35, 15×2+15=45, 20×2+15=55, and 25×2+15=65.

11. Answer: A
Explanation: Number of cups =2,750
Each box has 25 cups.
Number of boxes Joseph used = 2,750÷25= 110

12. Answer: D
Explanation: The number 503 cannot be divided equally into three groups.
So, the number of cars in Mr. Michael's showroom =503

13. Answer: B
Explanation: Add 42 plus 30 and get 72. Then, divide 72 by 6, which equals 12 with a remainder of 0. Tom can make 12 full boxes of 6 shirts in each box.

ANSWERS AND EXPLANATIONS

14. Answer: C
Explanation: The list of composite numbers are 48, 51, 62, and 81

15. Answer: D
Explanation: The list that contains two prime numbers is 28, 36, 47, 51, 53, and 63.

16. Answer: B
Explanation: 8×13=104

17. Answer: A
Explanation: 4 & 40 is not a factor pair of 120.

18. Answer: C
Explanation: 5, 9 are prime numbers.
9−5=4

19. Answer: D
Explanation: Number of flower pots =12,000
Each row has 250 flower pots.
Number of rows =12,000÷250=48

20. Answer: B
Explanation: Nora's age =x
x×500+700=5,700

$$x = \frac{5{,}700 - 700}{500}$$

$$x = \frac{5{,}000}{500}$$

x = 10

4. FRACTIONS

4.1 Equivalent Fraction Using an Area Model

1. Answer: C
Explanation: The fraction 60÷300 is equivalent to 6÷30.
6÷30=1÷5 and 60÷300=1÷5

2. Answer: A
Explanation: The fraction of the straws that are violet is 16 straws out of 40 or 16÷40
The fraction 2÷5 is equivalent to 16÷40 because 2÷5 × 8÷8=16÷40 and 8÷8 is equal to 1.

3. Answer: B
Explanation: The shaded portion represents the fraction 8÷12 which can be rewritten as the fraction 23 . The fractions 2÷3, 16÷24, 24÷36 are equivalent to 8÷2

4. Answer: D
Explanation: The fraction 9÷4 is the largest fraction.

5. Answer: 70÷100 9÷20 2÷25 4÷10
Explanation: The fractions can be compared as fractions out of 100
4÷10=40÷100,
2÷25=8÷100, and
9÷20=45÷100

ANSWERS AND EXPLANATIONS

6. Answer: B
Explanation: $3 \div 4 = 9 \div 12$

7. Answer: C
Explanation: $1 \div 2 = 5 \div 10$

8. Answer: A
Explanation: $8 \div 16 = 1 \div 2$

9. Answer: D
Explanation: $10 \div 15 = 20 \div 30$

10. Answer: C
Explanation: $5 \div 12 = 15 \div 36$

11. Answer: 6
Explanation: $3 \div 4 = k \div 8$
$k \times 4 = 8 \times 3$
$k = 6$

12. Answer: D
Explanation: Emily ate 3 slices of apples. Ella ate 6 slices of apples.

13. Answer: A
Explanation: This circle is divided into 12 equal parts, 4 of which are shaded. The shaded part represents $4 \div 12$.
The circle can be seen as being split into six parts, with two of the parts being shaded. Thus, $2 \div 6 = 4 \div 12$

14. Answer: C
Explanation: This circle is divided into 24 equal parts, 16 of which are shaded. The shaded part represents $16 \div 24$.
The circle can be seen as being split into twelve parts, with eight of the parts being shaded.
Thus, $8 \div 12 = 16 \div 24$

15. Answer: B
Explanation: $1 \div 4 = 2 \div 8$
The first circle is divided into 4 equal parts, 1 of which is shaded.
The shaded part represents 14
The second circle is divided into 8 equal parts, 2 of which are shaded.
The shaded part represents 28
The same amount in both circles is shaded. Therefore, $14 = 28$

4.2 Equivalent Fraction Using a Length Fraction Model

1. Answer: A
Explanation:
$5 \div 9 = (5 \times 2) \div (9 \times 2) = (5 \times 3) \div (9 \times 3) = (5 \times 4) \div (9 \times 4)$ therefore,
$5 \div 9 = 10 \div 18 = 15 \div 27 = 20 \div 36$

2. Answer: C
Explanation: $26 \div 52 = (26 \div 2) \div (52 \div 2)$
$(26 \div 13) \div (52 \div 13)$ therefore,
$26 \div 52 = 13 \div 26 = 2 \div 4$

3. Answer:

Explanation:
$(10 \div 2) \div (12 \div 2) = 5 \div 6$

ANSWERS AND EXPLANATIONS

4. Answer:

```
+--+--+--+--+--+--+--●--+--+--+
0  1/10 2/10 3/10 4/10 5/10 6/10 7/10 8/10 9/10  1
```

Explanation:
$(7 \times 10) \div (10 \times 10) = 70 \div 100$

5. Answer: C
Explanation: $36 \div 54 = 2 \div 3$
36 out of 54 students are girls. Or, it is the same as $36 \div 54$
Express $36 \div 54$ in its simplest form.
The factors of 36 are: 1, 2, 3, 4, 6, 9, 12, 18, 36.
The factors of 40 are: 1, 2, 3, 6, 9, 18, 28, 54.
$(36 \div 18) \div (54 \div 18) = 23$

6. Answer: B
Explanation: 20 of 120 is same as 20120
Or $20 \div 120 = (20 \div 20) \div (120 \div 20) = 1 \div 6$

7. Answer: A
Explanation: $15 \div 9 = (15 \div 3) \div (9 \div 3) = 5 \div 3$

8. Answer: D
Explanation: 12 of 48 is the same as $12 \div 48$ or $12 \div 48 = (12 \div 12) \div (48 \div 12) = 1 \div 4$

9. Answer: C
Explanation: $7 \div 56 = (7 \div 7) \div (56 \div 7) = 1 \div 8$

10. Answer: 1
Explanation: $5 \div 45 = 1 \div 9$
We know that $45 \div 5 = 9$
Therefore $5 \div 45 = (5 \div 5) \div (45 \div 5) = 1 \div 9$

11. Answer: B
Explanation: $4 \div 7 = (4 \times 3) \div (7 \times 3) = 12 \div 21$

12. Answer: C
Explanation: $2 \div 3 = (2 \times 16) \div (3 \times 16) = 32 \div 48$

13. Answer: $4 \div 9$
Explanation: $80 \div 180 = 4 \div 9$
80 out of 180 is $80 \div 180$
$(80 \div 10) \div (180 \div 10) = 8 \div 18 = (8 \div 2) \div (18 \div 2) = 4 \div 9$
Hence, the lowest equivalent form of $80 \div 180$ is $4 \div 9$

14. Answer: D
Explanation: We can write:
$43 = (4 \times 2) \div (3 \times 2) = (4 \times 3) \div (3 \times 3) = (4 \times 4) \div (3 \times 4)$ therefore
$4 \div 3 = 8 \div 6 = 12 \div 9 = 16 \div 12$
$7 \div 6 = (7 \times 2) \div (6 \times 2) = (7 \times 3) \div (6 \times 3)$
therefore, $7 \div 6 = 14 \div 12 = 21 \div 18$
$12 \div 10 = (12 \div 2) \div (10 \div 2)$ therefore, $12 \div 10 = 6 \div 5$
So, we will have $4 \div 3 = 6 \div 12$, $7 \div 6 = 21 \div 28$, and $12 \div 10 = 6 \div 5$

15. Answer: $34 \div 102$, $1 \div 3$
Explanation: $17 \div 51 = (17 \times 2) \div (51 \times 2)$
therefore, $17 \div 51 = 34 \div 102$
$17 \div 51 = (17 \div 17) \div (51 \div 17)$ therefore, $17 \div 51 = 1 \div 3$

ANSWERS AND EXPLANATIONS

4.3 Add and Subtract Fraction

1. Answer: B
Explanation: The first model has 4 parts shaded out of 10 parts, which is 4÷10. The second model has 6 parts shaded out of 10 parts, which is 6÷10. The model adds the two fractions: 4÷10+6÷10.

2. Answer: D
Explanation: When adding fractions with the same denominator, add the numerators, but do not change the denominators.
$(\frac{1}{7} + \frac{1}{7} + \frac{1}{7} + \frac{1}{7} + \frac{1}{7})$

3. Answer: A
Explanation: Add the numerators. $\frac{3}{9} + \frac{5}{9} = \frac{8}{9}$. Do not change the denominators.

4. Answer: C
Explanation: Subtract the numerators. $\frac{10}{12} - \frac{7}{12} = \frac{3}{12}$. Do not change the denominators.

5. Answer: D
Explanation: Adding the numerators, $\frac{4}{3} + \frac{8}{3} = \frac{12}{3}$, shows that the resulting fraction has a numerator that is larger than the denominators, which means the fraction is larger than 3.

6. Answer: B
Explanation: The whole watermelon cut into 12 pieces is 12÷12. If Emma ate 8÷12 of the watermelons, subtract to find the remaining part $\frac{12}{12} - \frac{8}{12} = \frac{4}{12}$

7. Answer: A
Explanation: Using the given information, 2÷7 of the T-shirts are black. Since there are 49 T-shirts, this means that 14 T-shirts were black.

8. Answer: C
Explanation: 4÷5×20=16

9. Answer: B
Explanation: $\frac{3}{13} + \frac{5}{13} + \frac{8}{13} = \frac{16}{13}$

10. Answer: C
Explanation: $\frac{12}{30} + \frac{16}{30} = \frac{28}{30}$

11. Answer: A
Explanation: $\frac{10}{80} = \frac{k}{40}$
k×80=40×10
k=5

12. Answer: B
Explanation: Max mixed $\frac{3}{13}$ cups of carrot juice and $\frac{7}{13}$ cups of beetroot juice. Therefore, he has have $\frac{3}{13} + \frac{7}{13} = \frac{10}{13}$ cups of juice.

ANSWERS AND EXPLANATIONS

13. Answer: C

Explanation: The two parts have a total amount of $\frac{6}{12} + \frac{5}{12} = \frac{11}{12}$
$1 - \frac{11}{12} = \frac{12}{12} - \frac{11}{12} = \frac{1}{12}$
$\frac{1}{12}$ part of Kelly's mix is pink.

14. Answer: D

Explanation: William cut
$\frac{7}{10} + \frac{77}{100} = \frac{70}{100} + \frac{77}{100} = \frac{147}{100}$
of the vegetables on both days.

15. Answer: B

Explanation: $t + \frac{17}{29} = \frac{22}{29}$
$t = \frac{22}{29} - \frac{17}{29} = \frac{5}{29}$

4.4 Mixed Numbers

---

1. Answer: A

Explanation: Change the mixed numbers into improper fractions. First rewrite 5 as a fraction with the same denominator as the fraction within the mixed number. Then add: $\frac{10}{2} + \frac{3}{2} = \frac{13}{2}$
Next, rewrite 4 as a fraction with the same denominator. Then, add
$\frac{8}{2} + \frac{3}{2} = \frac{11}{2}$. Add the results.

2. Answer: C

Explanation: First, let's convert the mixed numbers to improper fractions:
$4\frac{5}{3} = \frac{17}{3}$ Next, let's add the two
$3\frac{7}{3} = \frac{16}{3}$ fractions together:
$\frac{17}{3} + \frac{16}{3} = \frac{33}{3} = 11$

Now, to determine the number of groups of $\frac{1}{3}$ hour, we divide 11 by $\frac{1}{3}$:
$11 \div \frac{1}{3} = 11 \times \frac{3}{1} = 33$

3. Answer: B

Explanation: The mixed number $8\frac{4}{6}$ can be rewritten as an improper fraction of for 52 days.

4. Answer: D

Explanation:
$14\frac{11}{35} + 17\frac{8}{35} = 14 + 17 + \frac{4}{6} + \frac{4}{6} = 31\frac{19}{35}$

5. Answer: C

Explanation:
$27\frac{9}{18} - 22\frac{3}{18} = 27 - 22 + \frac{9}{18} - \frac{3}{18} = 5\frac{6}{18}$

6. Answer: A

Explanation:
$t + 12\frac{13}{24} = 19\frac{16}{24}$
$t = 19\frac{16}{24} - 12\frac{13}{24} = 19 - 12 + \frac{16}{24} - \frac{13}{24} = 7\frac{3}{24}$

7. Answer: B

Explanation: $36\frac{18}{41} - t = 30\frac{18}{41}$
$t = 36\frac{18}{41} - 30\frac{18}{41} = 36 - 30 + \frac{18}{41} - \frac{18}{41} = 6$

8. Answer: D

Explanation:
$14\frac{35}{200} - 9\frac{25}{200} = 14 - 9 + \frac{35}{200} - \frac{25}{200} =$
$5\frac{10}{200} = 5\frac{1}{20}$

312

ANSWERS AND EXPLANATIONS

9. Answer: C
Explanation:
$15\frac{18}{50} - 4\frac{9}{50} = 15 - 4 + \frac{18}{50} - \frac{9}{50} = 11\frac{9}{50}$

10. Answer: B
Explanation:
$5\frac{8}{2} - 3\frac{5}{2} = 5 - 3 + \frac{8}{2} - \frac{5}{2} = 2\frac{3}{2}$

11. Answer: D
Explanation:
$7\frac{6}{10} - 9\frac{7}{10} = 7 + 9 + \frac{6}{10} + \frac{7}{10} = 16\frac{13}{10}$

12. Answer: A
Explanation:
The mixed number $31\frac{2}{6}$ can be rewritten as an improper fraction of $\frac{188}{6}$. This fraction contains 188 groups of $\frac{1}{6}$. This means Jerry has worked 16 hours for 188 days.

13. Answer: C
Explanation:
$70\frac{8}{55} + 49\frac{13}{55} + 26\frac{10}{55} = 70 + 49 + 26 + \frac{8}{55} + \frac{13}{55} + \frac{10}{55} = 145\frac{31}{55}$

14. Answer: B
Explanation:
$10\frac{2}{70} + x = 15\frac{26}{70}$
$x = 15\frac{26}{70} - 10\frac{2}{70} = 15 - 10 + \frac{26}{70} - \frac{2}{70} = 5\frac{24}{70}$

15. Answer: D
Explanation:
The mixed number $20\frac{7}{4}$ can be rewritten as an improper fraction of $\frac{87}{4}$. This fraction contains 87 groups of $\frac{1}{4}$. This means Ian has worked $\frac{1}{4}$ hours for 87 days.

ANSWERS AND EXPLANATIONS

4.5 Chapter Review

1. Answer: A
Explanation: The fraction 7÷49 can be simplified into equivalent fraction 1÷7 because 7 in the numerator and 49 in the denominator have a common factor 7 which can be divided out
(7÷49)÷(7÷7)=(1÷7)

2. Answer: C
Explanation: Represent the original amount of apple juice with the fraction 12÷12. The difference between 5÷12 and 4÷12 is 1÷12. Then 5÷12 is equal to 1÷12 added to itself 5 times.

3. Answer: B
Explanation: Multiply 1÷5 by 15 or divide 15 by 5 to get 3. Multiply 1÷7 by 28 or divide 28 by 7 to get 4.

4. Answer: C
Explanation: Multiply 45 by 15 and multiply 9÷90 by 15÷1 to get 135÷90. Combine the two results into a new mixed fraction $675\frac{135}{90}$.

5. Answer: 90÷200
Explanation: To have an equivalent fraction with a denominator of 200, multiply the numerator and denominator by 10

6. Answer: 80÷100
Explanation:
60÷100+2÷10=60÷100+20÷100=80÷100

7. Answer: D
Explanation: 8÷12=24÷36

8. Answer: 38÷228, 1÷6
Explanation: 19÷114=(19×2)÷(114×2) therefore, 19÷114=38/228
19÷114=(19÷19)÷(114÷19) therefore, 19÷114=1÷6

9. Answer: B
Explanation: The first model has 12 parts shaded out of 18 parts, which is 12÷18. The second model has 9 parts shaded out of 18 parts, which is 9÷18. The model adds the two fractions: 12÷18+9÷18.

10. Answer: A
Explanation: (6÷8)×(1÷2)
Number of shaded parts=6
Number of unshaded parts =10
Total=6+10=16

$$\frac{6}{16} = \frac{6}{8} \times \frac{1}{2}$$

11. Answer: B
Explanation:

$$3\frac{2}{3} \times \frac{14}{33} - \frac{1}{9} = \frac{11}{3} \times \frac{14}{33} - \frac{1}{9} = \frac{1}{3} \times \frac{14}{3} - \frac{1}{9}$$

$$\frac{14}{9} - \frac{1}{9} = \frac{13}{9} = 1\frac{4}{9}$$

ANSWERS AND EXPLANATIONS

12. Answer: C
Explanation:
$9 \div 4\frac{7}{5} + \frac{2}{3} = \frac{9}{1} \div \frac{27}{5} + \frac{2}{3} = \frac{9}{1} \times \frac{5}{27} + \frac{2}{3}$
$= \frac{5}{3} + \frac{2}{3} = \frac{7}{3} = 2\frac{1}{3}$

13. Answer: B
Explanation: Change the mixed numbers into improper fractions. First rewrite 7 as a fraction with the same denominator as the fraction within the mixed number. Then add: $\frac{21}{3} \times \frac{4}{3} = \frac{25}{3}$
Next, rewrite 8 as a fraction with the same denominator. Then, add
$\frac{21}{3} \times \frac{2}{3} = \frac{26}{3}$
Add the results.

14. Answer: D
Explanation:
$2\frac{2}{5} \times 3\frac{8}{4} = \frac{12}{5} \times \frac{20}{4} = 12$ miles

15. Answer: A
Explanation:
$15\frac{6}{2} \div \frac{6}{14} = \frac{36}{2} \times \frac{14}{6} = 42$ miles

16. Answer: C
Explanation:
$5\frac{8}{4} \div \frac{7}{12} = \frac{28}{4} \times \frac{12}{7} = 12$
$4 \div b = 12$
$b = \frac{12}{4}$
$b = 3$

17. Answer: B
Explanation:
One whole wheat loaf of bread: $3\frac{7}{10}$ cups
Six whole wheat bread loaves:
$3\frac{7}{10} \times 6 = \frac{37}{10} \times \frac{6}{1} = \frac{111}{5} = 22\frac{1}{5}$

18. Answer: C
Explanation:
$\frac{1}{6} - \frac{1}{8} = \frac{8}{48} - \frac{6}{48} = \frac{2}{48} = \frac{1}{24}$
$\frac{8}{12} \div \frac{1}{24} = \frac{8}{12} \times \frac{24}{1} = 16$
$19 - 16 = 3$
$7 \div 3 = 2\frac{1}{3}$
$8 - 2\frac{1}{3} = (7 - 2) + (1 - \frac{1}{3}) = 5 + \frac{2}{3} = 5\frac{2}{3}$

19. Answer: A
Explanation:
$7 \div 4\frac{5}{4} + \frac{1}{3} = \frac{7}{1} \div \frac{21}{4} + \frac{1}{3} = \frac{7}{1} \times \frac{4}{21} + \frac{1}{3} =$
$\frac{4}{3} + \frac{1}{3} = \frac{5}{3} = 1\frac{2}{3}$

20. Answer: 18 cm
Explanation:
Base = $3\frac{7}{10}$ cm
$\frac{1}{2} \times$ Base $= \frac{1}{2} \times 3\frac{6}{3} = \frac{1}{2} \times \frac{15}{3} = \frac{15}{6} = \frac{5}{3}$ cm
Height = (Area) $\div (\frac{1}{2} \times$ Base$)$
$= \frac{30}{1} \div \frac{5}{3} = \frac{30}{1} \times \frac{3}{5} = 18$ cm

ANSWERS AND EXPLANATIONS

5. CONVERSIONS

5.1 Time Across the Hours

1. Answer: A
Explanation: From 10 a.m. to noon, it is 2 hours. Then, from noon to 1:30 p.m., it is 1 hour and 30 minutes. Add the two quantities of time together.

2. Answer: B
Explanation: 1 hour=60 minutes
4 hours=4×60=240 minutes

3. Answer: D
Explanation: 2 hours 30 minutes+5 hours 40 minutes=8 hours 10 minutes

4. Answer: C
Explanation: 1 sec=$\frac{1}{60}$ min
720 seconds=$\frac{720}{60}$=12 minutes

5. Answer: B
Explanation: 1 hour=60 minutes
Kelly danced =3 hours=3×60= 180 minutes
Lily danced =132 minutes
Kelly spent more time datncing.

6. Answer: C
Explanation: 11:55 p.m.–7:30 p.m. 4 hours 25 minutes

7. Answer: A
Explanation: From 12:15 a.m to 1:00 a.m is 45 minutes

8. Answer: D
Explanation: 2:05 p.m.+45 minutes= 2:50 p.m.

9. Answer: B
Explanation: From 1:00 p.m. to 3:00 p.m. is 2 hours.
From 3:00 p.m. to 3:30 p.m. is 30 minutes
2 hours+30 minutes=2 hours 30 minutes

10. Answer: C
Explanation: 9:30 a.m+60 minutes=10:30 a.m

11. Answer: A
Explanation: From 2:15 p.m. to 4:00 p.m. is 1 hour 45 minutes
From 4:00 p.m. to 4:47 p.m. is 47 minutes
1 hour+45 minutes+47 minutes=2 hours 32 minutes

12. Answer: C
Explanation: 8:30 p.m.+60 minutes= 9:30 p.m.

13. Answer: B
Explanation:
11:25 a.m.+2 hours 10 minutes=1:35 p.m.

14. Answer: C
Explanation: 1 year=365 days
365×40=$\frac{14,600}{60}$=243.3 hours

15. Answer: D
Explanation: From 9:00 p.m. to 10:00 p.m. is 1 hour

ANSWERS AND EXPLANATIONS

1 hour=60 minutes
60 minutes−20 minutes=40 minutes
Therefore, the TV show lasted 40 minutes.

5.2 Relating Conversions to Place Value

1. Answer: B
Explanation: 1 km=1,000 m
7 km=7×1000=7,000 m
7,000 m+5,000 m=12,000 m

2. Answer: A
Explanation: 1 cm=10 mm
500 mm×1÷10=50 cm

3. Answer: D
Explanation: 5×85 cm=425 cm
425×10=4,250 mm

4. Answer: C
Explanation: 1 km=1,000 m
6×4 km=24 km
24×1,000=24,000 m

5. Answer: B
Explanation: Subtract the difference to find the length.
105 cm−55 cm=50 cm
There are 10 mm in 1 cm
50×10=500 mm

6. Answer: A
Explanation: 16 kg÷8=2 kg
1 kg=1000 grams
2×1,000=2,000 g

7. Answer: B
Explanation: Multiply to get a total of 6,000 pencils: 6,000×2 g=12,000 g
1 kg=1000 grams
12,000÷1,000=12
6,000 pencils have a total mass of 12 kg.

8. Answer: C
Explanation: There are 15×20=300 carrots
300×13 g=3,900 g
1 kg=1,000 grams
3,000÷1,000=3
3,900 g=3000 g+900 g=3 kg 900 g

9. Answer: B
Explanation: 1 kg=1,000 grams
3 kg=3×1,000=3,000 g
3,000 g÷75g=40
There are 40 cupcakes in the box.

10. Answer: D
Explanation: 180×1,000=180000 ml
110 L 200 ml=110200 ml
180000−110200=69800 ml=69 L 800 ml

11. Answer: C
Explanation: 35 kg 250 g+38 kg 120 g=73 kg 370 g=73370 g
1 g=1,000 mg
73,370 g=73,370 g×1,000=73,370,000 mg

12. Answer: A
Explanation: 11L=11000 ml
11000 ml+710 ml=11710 ml

ANSWERS AND EXPLANATIONS

13. Answer: B
Explanation: 10×2 glasses=20 glasses
20×150 ml=3000 ml
3000 ml=3 L
4 L>3 L

14. Answer: C
Explanation: Dylan's weight in grams:
42 kg 800 g=42,800 g
Felipe's weight in grams:
42,800 g−4,980 g=37,820 g
Quentin's weight in grams:
37,820g−2,400 g=35,420 g

15. Answer: D
Explanation: 5 m 20 cm=520 cm
Second curtain =520 cm−56 cm=464 cm
Third curtain=464 cm+157 cm=621 cm
621 cm−520 cm=101 cm
1 cm=10 mm
101×10=1010 mm

5.3 Centimetres and Meters

1. Answer: 32 cm 9 mm
Explanation: 24 cm 8 mm+8 cm 1 mm=32 cm 9mm

2. Answer: 7m 65 cm
Explanation: 2 m 45 cm+5 m 20 cm= 7 m 65 cm

3. Answer: 3 m 88 cm
Explanation: 4 m−12 cm=400 cm−12 cm =388 cm=3 m 88 cm

4. Answer: C
Explanation: 34 cm=34×10=340 mm

5. Answer: D
Explanation: 7 m=7×1000=7000 mm

6. Answer: B
Explanation: 18 km 30 m÷2=9 km 15 m=9015 m

7. Answer: C
Explanation: 2 m 54 cm=200 cm 540 mm=2,000 mm 540 mm=2,540 mm

8. Answer: A
Explanation: 5 cm=50 mm
92 mm+50 mm=142 mm

9. Answer: B
Explanation: 6 m 64 cm=664 cm=6640 mm

10. Answer: D
Explanation: 30 cm=30×10=300 mm

11. Answer: A
Explanation: 12 km=12,000 m

12. Answer: C
Explanation: 5 m×100=500 cm

13. Answer: A
Explanation: 580 cm=580×10=5,800 mm
5,800>5,160

ANSWERS AND EXPLANATIONS

14. Answer: B
Explanation: 4 km=4×1,000=4,000 m
4000 m>1800 m

15. Answer: 8 km
Explanation: 8 km=8×1,000=8,000 m=8,000×1,000=8,000,000 mm
8,000,000 mm>860,000 mm
Therefore, 8 km>860,000 mm

5.4 Grams and Kilograms

1. Answer: B
Explanation: 6 g×12=72 g
72 g×40=2,880 g=2 kg 880 g

2. Answer: D
Explanation: 450 g+900 g=1,350 g=1 kg 350 g

3. Answer: C
Explanation: 2 kg 400 g=2,400 g
2,400 g−625 g=1,775 g

4. Answer: A
Explanation: 2 kg=2×1,000=2,000 g
The weight of a mouse is 2,000 g

5. Answer: B
Explanation: The weight of the mirror =14 kg 275 g=14,275 g

6. Answer: A
Explanation: 60 g=60×1,000=60,000 mg

7. Answer: C
Explanation: 80×30 g=2,400 g
2,400 g÷1,000=2 kg 400 g

8. Answer: A
Explanation: The weight of a pumpkin=3,800 g
The weight of three cauliflowers = 1 kg×3=3 kg=3,000 g
So, 3,800 g>3,000 g

9. Answer: B
Explanation: 8,400 g=8 kg 400 g
8 kg 400 g<13 kg

10. Answer: A
Explanation: 62 kg 250 g−34 kg 170 g= 28 kg 80 g

11. Answer: D
Explanation: 2 g×250=500 g

12. Answer: B
Explanation: 10 kg 640 g÷8=10,640 g÷8=1,330 g
One pineapple weight is 1 kg 330 g

13. Answer: C
Explanation: 32 kg+16 kg 200 g=48 kg 200 g
The total weight of the apricots=48 kg 200 g

ANSWERS AND EXPLANATIONS

14. Answer: A
Explanation: 12 kg 120 g ÷ 5 = 12,120 g ÷ 5 = 2,424 g
One watermelon weight is 2 kg 424 g

15. Answer: C
Explanation: 7 kg 458 g = 7,458 g
7,458 g + 674 g = 8,132 g = 8 kg 132 g
Royce plucked 8 kg 132 g of mangoes.

5.5 Litres and Millilitres

1. Answer: No
Explanation: 5 L = 5,000 ml

2. Answer: 120,000 ml
Explanation: 48 L + 72 L = 120 L = 120,000 ml

3. Answer: 81 L
Explanation: 46,800 ml + 34,200 ml = 81,000 ml = 81 L

4. Answer: B
Explanation: 10 L = 10 × 1,000 = 10,000 ml

5. Answer: D
Explanation: 18 L = 18 × 1,000 = 18,000 ml
18,000 ÷ 250 = 72

6. Answer: A
Explanation: 4 L 500 ml − 3 L 300 ml = 4,500 ml − 3,300 ml = 1,200 ml
1,200 ml + 450 ml = 1,650 ml = 1 L 650 ml

7. Answer: C
Explanation: 3 L = 3,000 ml
$3,000 \times \dfrac{3}{9} = \dfrac{9,000}{9} = 1,000$ ml is used
3,000 − 1,000 = 2,000 ml is left
$2,000 \times \dfrac{1}{4} = 500$ ml is used
2,000 − 500 = 1,500 ml

8. Answer: 300 ml
Explanation: 2 L 700 ml = 2,700 ml
2,700 × 13 = 900 ml
2,400 − 900 = 1,500 ml
1,500 ÷ 5 = 300 ml

9. Answer: 25L 480 ml
Explanation: 24 L = 24,000 ml
24,000 + 1480 = 25,480 ml = 25 L 480 ml

10. Answer: B
Explanation: 250 ml × 9 = 2,250 ml = 2 L 250 ml

11. Answer: D
Explanation: 1,050 ml × 21 = 22,050 ml = 22 L 50 ml

12. Answer: C
Explanation: 5 L 980 ml = 5,980 ml
5,980 ÷ 260 = 23

13. Answer: A
Explanation: 1,720 ml + 3 L = 1,720 + 3,000 = 4,720 ml
4,720 ml ÷ 16 = 295 ml

ANSWERS AND EXPLANATIONS

14. Answer: C
Explanation: 1 L=1,000 ml
So 1/5 of 1,000 ml=200 ml
Therefore 10,000÷200=50 bottles

15. Answer: B
Explanation: 92×320 ml=29,440 ml
2 L=2,000 ml
29,440÷2,000=14.72=15 (approximately)

- - - - - - - - - - - - - - -
5.6 Chapter Review
- - - - - - - - - - - - - - -

1. Answer: A
Explanation: Add the times to find the total elapsed time. Next, find the time that is 3 hours after 1:30 p.m. The soccer practice ends at 4: 30 p.m.

2. Answer: B
Explanation: Add the minutes to find the total time per day. Next, multiply 60 by 6. Then, convert it to hours and minutes. He spends 6 hours traveling to his job.

3. Answer: C
Explanation: 3 kg-1 kg 95 g=3,000-1,095=1,905 g

4. Answer: D
Explanation: 6×450 ml=2,700 ml
2,700÷5=540 ml

5. Answer: 1 L 600 ml
Explanation: 10×160 ml=1,600 ml= 1 L 600 ml

6. Answer: B
Explanation: 3 hours 25 minutes+1 hours 50 minutes=5 hours 15 minutes, that is 5:15 p.m.

7. Answer: C
Explanation: 6 hours 50 minutes-3 hours 30 minutes=3 hourst 20 minutes, that is 3:20 p.m.

8. Answer: A
Explanation: 4 days, 30 hours, 1260 minutes, 700 seconds

9. Answer: D
Explanation: 1 sec=$\frac{1}{60}$ min
3,840 seconds=$\frac{3,840}{60}$=64 minutes

10. Answer: B
Explanation: From 2:35 p.m. to 5:00 p.m. is 2 hour 25 minutes
From 5:00 p.m. to 5:50 p.m. is 50 minutes
2 hour+25 minutes+50 minutes=3 hours 15 minutes

11. Answer: A
Explanation: $\frac{48 \text{ kg}}{12}$=4 kg
1 kg=1000 grams
4×1,000=4,000 g

12. Answer: C
Explanation: 12 kg 150 g+18 kg 160 g=30 kg 310 g=30,310 g
1 g=1,000 mg
30,310 g=30,310 g×1,000=30,310,000 mg

ANSWERS AND EXPLANATIONS

13. Answer: A
Explanation: First ladder =6 m 80 cm=680 cm
Second ladder =680 cm−70 cm=610 cm
Third ladder=610 cm+184 cm=794 cm
794 cm−610 cm=184 cm
1 cm=10 mm
184×10=1,840 mm

14. Answer: C
Explanation: 58 cm=58×10=580 mm

15. Answer: B
Explanation: 11 m ×100=1,100 cm

16. Answer: B
Explanation: 290 cm=290×10=2,900 mm
2,900>1,240

17. Answer: A
Explanation: 45 kg=45×1,000=45,000 g
The weight of a baby horse is 45,000 g

18. Answer: C
Explanation: 16 kg 140 g÷10=16,140 g÷10=1,614 g
One red cabbage weight is 1 kg 614 g

19. Answer: A
Explanation: 5 L 100 ml−2 L 700 ml=5,100 ml−2,700 ml=2,400 ml
2,400 ml+670 ml=3,070 ml=3 L 70 ml

20. Answer: B
Explanation: 4 L 550 ml=4,550 ml
4,550÷350=13

6. GEOMETRIC MEASUREMENT

6.1 Area of Rectilinear

1. Answer: A
Explanation: Area of a larger square is 10×10=100 square units
Area of a small square is 2×2=4 square units. Area of the unshared shape is 100−4=96 square units

2. Answer: D
Explanation: The shape is divided into three rectangles. Their areas are:
2×2=4 square centimeters
4×4=16 square centimeters
2×2=4 square centimeters
The area of the shape is 4+16+4=24 square centimeters.

3. Answer: C
Explanation: There are 3+5=8 small squares. A=188=144 square units.

4. Answer: 28 square inches
Explanation: The whole rectangle is made up of two non-overlapping parts, the shaded part and the unshaded part. The area of the shaded region is 4×3 square inches. The area of the unshaded region is 4×4 square inches.
The area of the rectangle is the sum of the areas of those two parts.
That is, (4×3+4×4) square inches.

ANSWERS AND EXPLANATIONS

Using the distributive property of multiplication over addition, we obtain $4\times3+4\times4=4\times(3+4)=4\times7=28$ square inches.

5. Answer: B
Explanation: The shape is divided into two rectangles. Their areas are:
$2\times14=28$ square meters
$10\times4=40$ square meters
The area of the shape is $28+40=68$ square meters.

6. Answers: A and C
Explanation: Cleo needs the area of the floor to know how many tiles to buy. There are 4 rows and 7 columns. Therefore, Area is $4\times7=7\times4$.
So, A and C are the correct options.

7. Answer: 4×6 square inches
Explanation: There are 4 rows of squares, and each row contains 6 unshaded squares. The area of the unshaded region is 4×6 square inches.

8. Answer: 116 square units
Explanation: The area of a larger rectangle is $16\times8=128$ square units. The area of a small rectangle is $4\times3=12$ square units.
The area of the unshared shape is $128-12=116$ square units.

9. Answer: D
Explanation: The shape is divided into three rectangles. Their areas are:
$2\times3=6$ square centimeters
$2\times6=12$ square centimeters
$2\times3=6$ square centimeters
The area of the shape is $6+12+6=24$ square centimeters.

10. Answer: A
Explanation: $A=7\times4=28$ cm^2.

11. Answer: D
Explanation: The shape is divided into three rectangles. Their areas are:
$2\times8=16$ square meters
$10\times2=20$ square meters
The area of the shape is $16+20=36$ square meters.

12. Answer: B
Explanation: $15\times3=45$ cm
$4\times3=12$ cm
$A=45\times12=540$ cm^2

13. Answer: C
Explanation: $4\times10=40$
$5\times4=20$
$40+20=60$
The area of the garden is $60 m^2$

14. Answer: 17 square units.
Explanation: $2\times6+5=12+5=17$ square units.

ANSWERS AND EXPLANATIONS

15. Answer: A
Explanation: The shelf can be divided into two rectangular parts:
60 cm by 60 cm
120 cm by 60 cm
Their areas are, respectively,
60×60=3600 cm² and 120×60=7200 cm²
The total area of the shelf is
3600+7200=10800 cm²

6.2 Fixed Area – Varying Perimeter
---

1. Answer: D
Explanation: Wall of height 3m and length 14m.
Area of the wall, $Area_1$ = length × height
$Area_1$ =3×14=42 m2
Same area with new height:
Wall of height 6 m and length x m
$Area_2$ = length × height
42=x×6
x=7
So, the same bucket of paint will cover a length of 7 m.

2. Answer: 10 cm, 60 cm
Explanation: Area, A = w×l
Substitute l=20 and A=200 in the equation.
200=w×20
Divide both sides by 10.
w=$\frac{200}{20}$=10
Perimeter, P=2(l+w)
P=2(20+10)=60 cm

3. Answer: B
Explanation: Area of the rectangle,
A = length × width =12×3=36 cm²
So, the area of the square is also 36 cm²
side length × side length =36 cm²
Note that, 36=6×6
So, the length of each side of the square is 6 cm.

4. Answer: l=13 cm and P=46 cm
Explanation: Area, A = width×length
Substitute A=130 and w=10 in the equation.
130=10×l
l=$\frac{130}{10}$=13 cm

Perimeter, P =2×(width + length)
Substitute l=13 and w=10 in the equation.
P=2×(10+13)=2×23=46 cm

5. Answer: 144 cm²
Explanation: Width =4 cm
Length 4×9=36 cm
Area of the rectangle,
A = length × width =36×4=144 cm²
So, area of the square is also 144 cm²
Note that, 144=12×12
So, length of each side of the square is 12 cm [Since, length = breadth in a square].

6. Answer: B
Explanation: The area of a rectangle is given by A = width length.
40=l×5
l=$\frac{40}{5}$=8 ft.

324

ANSWERS AND EXPLANATIONS

7. Answer: D
Explanation: The area of a rectangle is given by A = width × length.
$63 = 7 \times w$
$w = \frac{63}{7} = 9$ yards.

8. Answer: B
Explanation: The area of a rectangle is given by A = width × length.
$42 = w \times 6$
$w = \frac{42}{6} = 7$ m
The perimeter of a rectangle is given by P = 2 × (width + length).
$P = 2(6+7) = 2 \times 13 = 26$ m.

9. Answer: D
Explanation: The area of a playground is given by A = width × length.
$920 = w \times 40$
$w = \frac{920}{40} = 23$ m
The perimeter P = 2(width + length).
$P = 2 \times (23+40) = 2 \times 63 = 126$ m

10. Answer: C
Explanation: The area of a rectangular shape is given by A = width × length.
$810 = w \times 90$
$w = \frac{810}{90} = 9$ cm
The perimeter of a rectangle is given by P = 2(width + length)
$P = 2(9+90) = 2 \times 99 = 198$ cm.

11. Answer: A
Explanation: The area of a rectangular garden is given by A = width × length.
$117 = 13 \times l$
$l = \frac{117}{13} = 9$ ft

12. Answer: A
Explanation: The area of the factory, $A_1 = 4000$ m²
Area of the warehouse, $A_2 = 4000$ m²
Warehouse length = l + 30 = 50 + 30 = 80 m
The area of warehouse,
A_2 = width × length
$4000 = w \times 80$
$w = \frac{4000}{80} = 50$ m
The perimeter of the warehouse,
$P_2 = 2(\text{width} + \text{length})$
$P_2 = 2 \times (50+80) = 2 \times 130 = 260$ m.

13. Answer: B
Explanation: The area of Liam's bedroom,
A = width × length
$A = 8 \times 2 = 16$ m²
The area of Owen's bedroom, A = The area of Liam's bedroom, A = 16 m²
Owen's bedroom, width = w + 2 = 2 + 2 = 4 m
The area of Owen's bedroom,
A = width × length
$16 = 4 \times l$
$l = \frac{16}{4} = 4$ m

ANSWERS AND EXPLANATIONS

14. Answer: 12 cm 6 cm and 9 cm ×8 cm
Explanation: The area of Noah's rectangle, A = width × length
$A = w \times 12 = 72 \text{ cm}^2$
$w = \frac{72}{12} = 6 \text{ cm}$

So, the dimensions of Noah's rectangle = 12 cm × 6 cm
The area of Logan's rectangle, A_2 = The area of Noah's rectangle = 72 cm²
Length of Logan's rectangle
= l − 4 = 12 − 4 = 8 cm
A = Width × Length.
72 = w × 8
w = 72⁄8 = 9 cm
So, the dimensions of Logan's rectangle = 9 cm × 8 cm

15. Answer: Shape with dimensions 3 cm 6 cm has the shortest perimeter; P = 18 cm
Explanation: Square area = 18 cm²
18 can be written as 2×9, 3×6
Perimeter P_1 = 9+9+2+2 = 22 cm
Perimeter P_2 = 3+3+6+6 = 18 cm
So, the shape with dimensions 3 cm × 6 cm has the shortest perimeter.

6.3 Fixed Perimeter − Varying Area
----

1. Answer: 6 cm and 24 cm².
Explanation: P = 2 × (l+w)
$w = \frac{P}{2} - l$
Substitute P = 20 and l = 4 in the equation.
w = 20⁄2 − 4; w = 10 − 4 = 6 cm
A = w × l = 6 × 4 = 24 cm²

2. Answer: 3 cm and 16 cm²
Explanation: Notice from the figure, l = 5 cm and b = 3 cm
Substitute, l = 5 cm and b = 3 cm in P = 2(l+w)
Perimeter of the rectangle, P = 2(3+5) = 16 cm
Given that perimeter of the square = perimeter of the rectangle
So, the perimeter of the square is also 16 cm.
Substitute P = 16 in P = 4a
4a = 16
a = 4
So, length of each side is 4 cm
Area of the square, A = 4 × 4 = 16 cm².

3. Answer: 54 m²
Explanation: P = 2l + 2w
Substitute P = 30 and l = 6 in the equation.
30 = 2 × 6 + 2w
30 = 12 + 2w
2w = 18
w = 9
A = 6 × 9 = 54 m²

4. Answer: B
Explanation: Let a be the width of the rectangular field. Then, 3a is the length of the rectangular field.
We can write:
2 × (3a + a) = 320
3a + a = 160
4a = 160
a = 40 m − width of the rectangular field
3a = 3 × 40 = 120 m − length of the rectangular field

ANSWERS AND EXPLANATIONS

5. Answer: A
Explanation: $P=2(l+w)$
Substitute P=60 and w=6 in the equation.
$60=2l+(2\times6)$
$60=2l+12$
$2l=48$
$l=24$
$A=24\times6=144$ ft²

6. Answer: C and D
Explanation: $P=8+3+x$
$18=8+3+x$
$x=18-8-3$
So, options C and D are correct.

7. Answer: C
Explanation: $P=2(l+w)$
Substitute l=8 and w=6 in the equation.
$P=2(8+6)$
$P=2\times14$
$P=28$ feet

8. Answer: x=99÷4
Explanation: The perimeter of a square is 99 inches. The length of its side is x inches.
$P=4x$
Therefore, $4x=99$
(Or) $x=99\div4$

9. Answer: A
Explanation: $P=2(l+w)$
Substitute P=22 and w=6 in the equation.
$22=2(l+6)$
$l+6=11$
$l=5$ inches.

10. Answer: B
Explanation: $P=2(l+w)$
Substitute P=180 and w=16 in the equation.
$180=2(l+16)$
$l+16=90$
$l=74$
$l=74$ ft
$A=74\times16=1,184$ ft²

11. Answer: 3200 m²
Explanation: If a is the width of the rectangular field, then, 2a is the length of the rectangular field.
Length of the fence is 240 m long, so the perimeter is 240 m.
We can write:
$2\times(2a+a)=240$
$2a+a=120$
$3a=120$
$a=40$ m – width of the rectangular field
$2a=2\times40=80$ m – length of the rectangular field
$A=40\times80=3200$ m²

12. Answer: 15120 m²
Explanation: Substitute l=200 and w=64 in $P=2(l+w)$
$P=528$ – perimeter of the basketball court
$l=200-20=180$ m – length of the basketball court. Basketball court has the same perimeter as a basketball court, then $528=2(180+w)$. $264=180+w$
$w=84$ - width of the basketball court
Area of the basketball court (A)
$=180\times84=15,120$ m²

ANSWERS AND EXPLANATIONS

13. Answer: B
Explanation: P=4a
We know, the perimeter is 40 inches.
So, 40=4a
a=40÷4
a=10 inches
Length of each side is doubled.
So, s=2×10=20 inches
New area, A=20×20=400 square inches.

14. Answer: A
Explanation: Substitute P=50 and w=5 in
P=2(l+w)
50=2(5+l)
Divide both sides by 2
25=5+l
Subtract 5 from both sides.
l=20

15. Answer: 3700 m² and 3520 m²
Explanation: P=2(l+w)
Substitute P=248 and l=74 in the equation.
248=2(74+w)
74+w=124
w=50 (width of the factory)
Substitute l=74 and w=50 in A=l×w
Area of the factory (A)=74×50=3700 m²
l=74+6=80 m – length of the warehouse
Warehouse has the same perimeter as factory,

then 248=2(80+w)
124=80+w
w=44m – width of the warehouse
Area of the warehouse A=80×44= 3520 m²

6.4 Perimeter and Area Formula for Rectangles

---

1. Answer: C and D
Explanation: Area of rectangle A=w×l
w=12+2=14 cm and l=8 cm
A=8×12+2=8×12+8×2.
Or A=8×14

2. Answer: 225 m² and 300
Explanation:
a) The area of a garden is 30×10=300 m²
The area of a patio is 300÷4=75 m²
The area of the lawn is 300−75=225 m²
b) The area of a tile is 50 cm×50 cm=2500 cm²
750000÷2500= 300

3. Answer: C
Explanation: The perimeter of triangle is
7+8+9=24 m
24÷4=6 m
P = (6+4)×2=20 m

4. Answer: P = 140 cm and S = 1200 cm²
Explanation: P = 140 cm
P = (30+40)×2=140 cm
S = 1200 cm²
S = 40×30=1200 cm².

ANSWERS AND EXPLANATIONS

5. Answer: Area of the room 1600 cm² and 300 carpet tiles are needed to cover the floor of the room
Explanation: 28÷2−8=6 m
6×8=48 m². The area of a tile is
40 cm×40 cm=1600 cm².
4,80,000÷1,600=300

6. Answer: B
Explanation: Perimeter of a square is greater than the perimeter of a rectangle. Perimeter of Square:
4a = 4×5=20 cm
P=2(l+w) = 6+5×2=22cm, 20 cm<22 cm

7. Answer: 24 cm
Explanation: P = (4+8)×2=24 cm.

8. Answer: 70 cm and 20 cm
Explanation: If the area of the flower bed is 1400 cm², then the dimensions of the flower bed can be: 70 cm and 20 cm.
P = (70+20)×2=180
So, the dimensions of the flower bed are 70 cm and 20 cm.

9. Answer: 3 m and 10 m.
Explanation: P=2(l+w)
If the area of the rectangular living room is 30 m², then the dimensions of the room can be:
3 m and 10 m
or 2 m and 15 m
or 5 m and 6 m
P = (3+10)×2=13×2=26
So, the dimensions of the living room are 3 m and 10 m.

10. Answer: B
Explanation: P=2(l+w)
(18+2)×2=40 feet

11. Answer: D
Explanation: The area of a floor is
6×24=144 m2
The area of a tile is 20×20=400 cm²
1440000 ÷ 400=3600
3600×4=14400

12. Answer: A
Explanation: The area of a swimming pool is 20×9=180 m²
30×19=570 m²
570 m²−180 m²=390 m²

13. Answer: B
Explanation: The area of the garden is 26 m×16 m=416 m²
The sidewalk is uniform in width 4m.
The area of the garden and sidewalk is (26+8)×(16+8)=34×24=816 m²
Therefore, the area of the sidewalk is 816 m²−416 m²=400 m²

14. Answer: D
Explanation: The area of a painting is 8 in × 20 in =160 in²
The frame is uniform with width 4 in.
The area of a painting without the frame is (8−4)in×(20−4) in = 64 in²
Therefore, the area of the frame is 160 in²−64 in²=96 in²

ANSWERS AND EXPLANATIONS

15. Answer: 144 minutes
Explanation: $(68+40) \times 2 = 216$
$216 \div 6 = 36$
$36 \times 4 = 144$ minutes

6.5 Chapter Review

1. Answer: 16 cm and 82 cm
Explanation: First find the width using the given length and area, using the formula:
width = $\dfrac{\text{area}}{\text{length}} = \dfrac{400}{25} = 16$ cm
Now find the perimeter using the formula: $2 \times (l + w) = 2 \times 25 + 16 = 82$ cm.

2. Answer: 5 cm and 40 cm²
Explanation: Perimeter, $P = 2(l+w)$
$w = \dfrac{p-l}{2}$
Substitute $P=26$ and $l=8$ in the equation.
$w = 13 - 8 = 5$ cm
Area of the rectangle, $A = l \times w$
$A = 8 \times 5 = 40$ cm²

3. Answer: B
Explanation: Area of a rectangle, $A = l \times w$
So, possible dimensions of a rectangle with area 10 square meters are:
10 m × 1 m (or) 5 m × 2 m
When the dimension is 10 m × 1 m,
Perimeter, $P = 2(10+1) = 2(11) = 22$ m
When the dimension is 5 m × 2 m,
Perimeter, $P = 2(5+2) = 2(7) = 14$ m
We need a rectangle with perimeter 22 m. So, the dimensions of the rectangle are 10 m × 1 m

4. Answer: D
Explanation: Notice that, 10 small squares can be drawn in the given figure. So, the area of the figure is 10 square units.

5. Answer: D
Explanation: Area of a square,
A = side
Area of the outer square, $A1 = 20 \times 2 = 400$ cm²
Area of a rectangle, A = length x width
Area of the inner rectangle,
$A2 = 10 \times 8 = 80$ cm²
Area of the shaded region = $A1 - A2 = 400 - 80 = 320$ cm²

6. Answer: A
Explanation: Area of a rectangle, $A = l \times w$
Substitute $A = 220$ and $w = 11$ in the equation.
$220 = l \times 11$
$l = \dfrac{220}{11} = 20$
So, the length of the walkway is 20 m.

7. Answer: B and D
Explanation: There are 4 clips along the length and 2 clips along the breadth.
So, perimeter = $4+2+4+2$
(or) perimeter = $(4 \times 2) + (2 \times 2)$

8. Answer: D
Explanation: Perimeter, $P = 2(l+w)$
So, width, $w = \dfrac{p-l}{2}$. Substitute $P=90$ and $l=30$ in the equation. $w = 45 - 30 = 15$ m
So, the dimensions of the field are 30 m

ANSWERS AND EXPLANATIONS

9. Answer: A
Explanation: The polygon has 12 sides. Each side is 7 inches.
So, the perimeter of the polygon
=12×7=84 inches

10. Answer: D
Explanation: For painting 4 portions Luke takes 20 minutes.
For painting 1 portion Luke takes $\frac{20}{4}=5$ minutes
Therefore, for painting 20 portions Luke will take 5×20=100 minutes

11. Answer: A
Explanation: Add all the outer side lengths.
There are 7 sides of length 3 cm and 1 side of length 9 cm.
Perimeter =(7×3)+(1×9)=21+9=30 cm

12. Answer: C
Explanation: Separate the figure into 2 squares and one rectangle. Find individual areas and add them up.

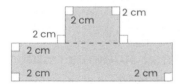

There is one square with side length 2 cm.
Area of the square, A1= (2×2)=4 cm²
There is one rectangle with length 6 cm and width 2 cm
Area of the rectangle, A2=6×2=12 cm²
Total area =A1+A2=4+12=16 cm²

13. Answer: Rectangles 2 and 4
Explanation: Add the number of squares in each rectangle.
Area of rectangle 1=20 square units
Area of rectangle 2=35 square units
Area of rectangle 3= 30 square units
Area of rectangle 4=35 square units
So, rectangles 2 and 4 have the same area.

14. Answer: 44 yards of fence and 96 square yards
Explanation: Length of the garden =16 yards; width of the garden=6 yards
Part (a):
Length of the fence is the perimeter of the garden.
Perimeter, P=2(l+w)
P=2(16+6)=44 yards
So, Mr. Isaac requires 44 yards of fence.
Part (b):
Area, A=l×w
A=16×6=96 square yards
So, the area of the garden is 96 square yards.

15. Answer: A
Explanation: Perimeter is the distance around the shape.
Perimeter, P = (8+3+2+8+3+2)cm

16. Answer: x=15 ft
Explanation: Given 272=16x+2
272=16x+32
272−32=16x
240=16x
x=15 ft

ANSWERS AND EXPLANATIONS

17. Answer: x=28 cm
Explanation: Given 104=218+6+x
$\frac{104}{2}=24+x$
52=24+x
x=28 cm

18. Answer: B
Explanation: Given Length 18 and Width=13+7
Area of the rectangle=18×(13+7)

19. Answer: D
Explanation: Given expression is 15x − 26
The side length of the square when x = 4
15×4−26=60−26=34.

20. Answer: C
Explanation: Area of the factory = 1320 m²;
Length of the factory = 60m
Area of the warehouse is the same as the factory.
So, area of the warehouse, A=1320 m²
Length of the warehouse, l=2×60=120 m
We know area, A=l×w
Substitute A=1320 and l=120 in the equation.
Width of the warehouse,
$w = \frac{A}{l} = \frac{1320}{120} = 11m$
Perimeter, P=2(l+w)
Substitute l=120 and w=11 in the equation.
P=2(120+11)
P=2(131)=262 m.

7. REPRESENT AND INTERPRET DATA

7.1 Bar Graphs and Frequency Tables

1. Answer: B
Explanation: From the given bar chart: most students chose Kiwi or Cherry.

2. Answer: A
Explanation: The quantities in the frequency column prtoperly correspond to the score range column.

3. Answer: C
Explanation: The highest recorded temperature: 14°C.

4. Answer: C
Explanation: The quantities in the frequency column properly correspond to the homerun range column.

5. Answer: D
Explanation: 20 people chose fish as their favorite meal.

6. Answer: B
Explanation: 21 students practiced for 30 minutes or less
12+4+5=12+9=21

ANSWERS AND EXPLANATIONS

7. Answer:

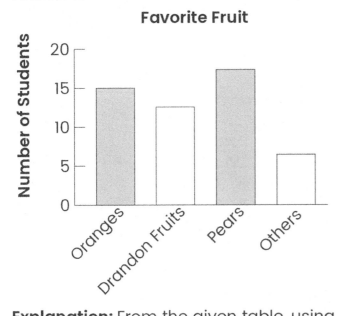

Explanation: From the given table, using the frequency value, we draw a bar graph

Favorite Fruits	Tally	Frequency
Grapes	|||| |||| |||| ||||	40
Orange	|||| |||| ||||	15
Drangon Fruit	|||| |||| ||	12
Pear	|||| |||| |||| ||	17
Others	||||		6

8. Answer: B
Explanation: 12+10-7-8=22-15=7

9. Answer:

Meal	Number of students
Burger	10
Pizza	7
Sandwich	8

Explanation:

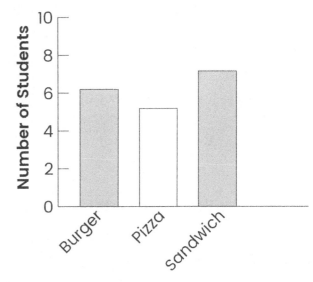

10. Answer: D
Explanation: 5+3+4-5-4=12-9=3

11. Answer: A
Explanation: Math: Grade 4- Grade 5 is 60-40=20

12. Answer:
Explanation: Total frequency is 31
Fraction of the absences occurred in December, January, February = $\frac{(3+3+3)}{31} = \frac{9}{31}$

ANSWERS AND EXPLANATIONS

13.Answer: A
Explanation: The dance event was attended by as many boys as girls (40+60=100)

14.Answer:
Explanation: Total frequency is 62
A fraction of the math club are freshmen and sophomores:
$$\frac{(12+20)}{62} = \frac{32}{62}$$

15.Answer: C
Explanation: The height of the column that corresponds to Yellow color chose 12 students.

7.2 Line Plots and Categorical vs. Numerical Data

1.Answer:

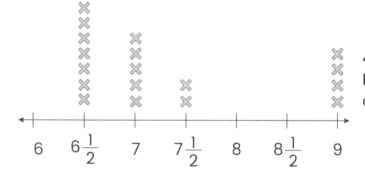

Explanation: Add 12 additional data points, from the table, to the line plot so the frequencies of each value match.

2.Answer:

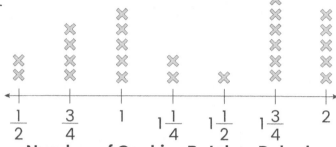

Number of Cookies Batches Baked

Explanation:
Add 10 additional data points, from the table, to the line plot so the frequencies of each value match.

3.Answer: D
Explanation: The line plot below matches the values presented in the table.

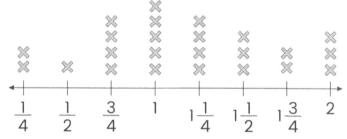

Time Spent Studying in hours

4.Answer:
Explanation: The data is quantitative and therefore numerical.

ANSWERS AND EXPLANATIONS

5. Answer:

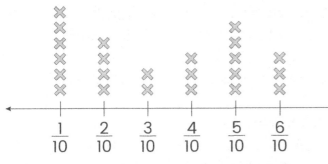

Length of Ribbon (in meters)

Explanation: Each value in the data table corresponds to a point on the line plot. The value in each row of the table is represented as a column in the line plot.

6. Answer: B
Explanation: The data is described as words and is categorical.

7. Answer: A
Explanation: The data is quantitative and therefore numerical.

8. Answer: A
Explanation: The data is quantitative and therefore numerical.

9. Answer: B
Explanation: The data is described as words and is categorical.

10. Answer: D
Explanation: Number of students who had $\frac{4}{10}$ of a bottle (or less) of water left
5+3+2+3=13

11. Answer: 5
Explanation: Determine the number of students by counting the marks to the left of the 1 cup position.

12. Answer:

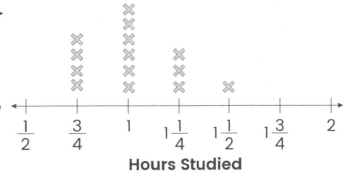

Hours Studied

Explanation:
Add 6 additional data points, from the table, to the line plot so the number of tallies match.

13. Answer:

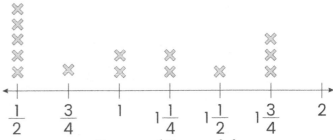

Hours of Excercising

Explanation: The line plot below matches the values present in the table.

14. Answer:

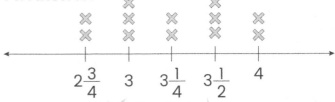

Length of Worms

ANSWERS AND EXPLANATIONS

Explanation:
The data set has 13 values that match the numbers on the number line. Plot the frequency of each value above the line in the appropriate place

15. Answer:

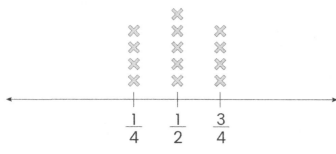

Cups of Sugar Per Day

Explanation:
The data set has 13 values that match the numbers on the number line. Plot the frequency of each value above the line in the appropriate place.

7.3 Chaper Review

1. Answer: $\frac{1}{2}$

Explanation: There are 5 grills which use $\frac{1}{2}$ gallons of fuel. This is the largest frequency shown on the line plot.

2. Answer: 6
Explanation: Count the number of data points to the right of $8\frac{1}{4}$ on the number line. There are 6 pencils longer than $8\frac{1}{4}$ inches.

3. Answer: 4
Explanation: Count the number of data points to the left of $\frac{3}{4}$. Four people used less than $\frac{3}{4}$ cup of paint.

4. Answer: 6
Explanation: Count the number of data points to the left of 1. Six people used less than 1 cup of paint.

5. Answer:

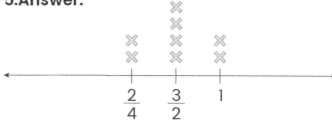

Amount of Red Paint in Tablespoons

Explanation:
The frequency of each value determines the number of points at each value on the line plot. For example, the value "$\frac{3}{4}$" occurs 5 times in the data set. It is marked with 5 points.

6. Answer:

Amount of Fuel Cans

ANSWERS AND EXPLANATIONS

Explanation:
The frequency of each value determines the number of points at each value on the line plot. For example, the value "$\frac{3}{4}$" occurs 2 times in the data set. It is marked with 2 points.

7. Answer:

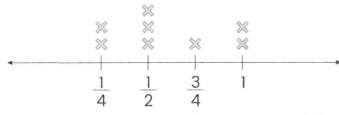
Hot Chocolate Mix (in tablespoons)

Explanation:
The frequency of each value determines the number of points at each value on the line plot. For example, the value "$\frac{2}{4}$" occurs 3 times in the data set. It is marked with 3 points.

8. Answer:

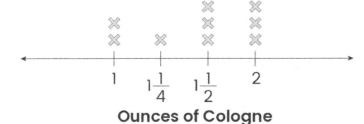
Ounces of Cologne

Explanation:
The frequency of each value determines the number of points at each value on the line plot. For example, the value "2" occurs 3 times in the data set. It is marked with 3 points.marked with 3 points.

9. Answer:

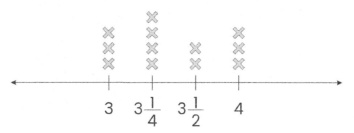

Explanation:
There are 12 data points in the line plot.

10. Answer:

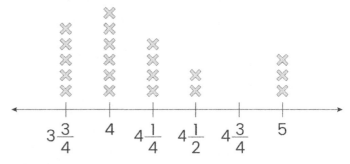

Explanation:
There are 20 data points in the line plot. One value is included in the scale ($4\frac{3}{4}$) but will not have any data points associated with it.

11. Answer: C
Explanation: 5+6-4-2=11-6=5

12. Answer: B
Explanation: From the given bar chart: most students chose "dog" or "parrot.

13. Answer: B
Explanation: 16-10=6

ANSWERS AND EXPLANATIONS

14. Answer: B
Explanation: $(1+2+4+2=9)$

15. Answer: B
Explanation: The data represents a quality and therefore is categorical.

16. Answer: A
Explanation: The data represents a quantity and therefore is numerical.

17. Answer: A
Explanation: The data represents a quantity and therefore is numerical.

18. Answer: B
Explanation: The data represents a quality and therefore is numerical.

19. Answer: D
Explanation:
$$\frac{(5+8)}{(5+8+4+2+3+2)} = \frac{13}{24}$$

20. Answer: B
Explanation: $3+6+2+3+2=16$.

8. MEASURING ANGLES

8.1 Types of Angles and Lines

1. Answer: A
Explanation: Angles measure between 0 and 360 degrees, a complete circle.

2. Answer: B
Explanation: Perpendicular lines meet at right angles. They make square corners.

3. Answer: B
Explanation: One pair of opposite sides will never cross. The shape has one pair of parallel sides.

4. Answer: A
Explanation: An acute angle measures less than 90 degrees.

5. Answer: A
Explanation: The opposite sides will never cross. So, these opposite sides are parallel.

6. Answer: C
Explanation: A right angle measures exactly 90 degrees.

7. Answer: A
Explanation: The triangle is a right triangle; one angle is exactly 90°.

8. Answer: B
Explanation: A ray has one endpoint and travels in one direction without stopping.

9. Answer: C
Explanation: A line segment has two endpoints.

10. Answer: B
Explanation: An obtuse angle measures more than 90 degrees.

ANSWERS AND EXPLANATIONS

11. Answer: C
Explanation: An angle is formed by two intersecting rays that share a common endpoint where they meet.

12. Answer: A
Explanation: A line goes without end in 2 directions.

13. Answer: A
Explanation: An acute angle measures less than 90 degrees.

14. Answer: B
Explanation: A line segment has two endpoints.

15. Answer: D
Explanation: Parallel lines stay the same distance apart. They will never intersect.

8.2 Measuring Angles Using a Protractor

1. Answer: B
Explanation: Determine the angle measure by reading the set of numbers on the protractor.

2. Answer: D
Explanation: Determine the angle measure by reading the set of numbers on the protractor.

3. Answer: A
Explanation: Angle OB is acute as it measures less than 90 degrees.

4. Answer: C
Explanation: The triangle is a right triangle; one angle is exactly 90°.

5. Answer: D
Explanation: The angle OD measures 27 degrees.

6. Answer: 15°
Explanation: Determine the angle measure by finding the difference between 95° and 80°.

7. Answer: A
Explanation: Based on the arc, the angle turns through the 0 degrees of the 360 degrees in a circle. The angle has a measurement in 0 degrees.

8. Answer: B
Explanation: The angle is greater than a right angle (90 degrees) and less than a straight angle (180 degrees). The best estimate is in 120 degrees.

9. Answer: 180°
Explanation: Calculate the total movement by multiplying: 6 × 30 = 180.

10. Answer: 20°
Explanation: Divide 1200 by 60 to determine the number of degrees moved.

ANSWERS AND EXPLANATIONS

11. Answer: 2
Explanation: There are 360 degrees in a circle. Divide 360 by 180.

12. Answer: B
Explanation: The angle measures over 90 degrees.

13. Answer: C
Explanation: Based on the shaded portion of the arc, the angle turns 180 degrees of the 360 degrees into a circle. The angle has a measurement of 180 degrees.

14. Answer: B
Explanation: The angle measures 46 degrees.

15. Answer: C
Explanation: 30-10=20 degrees

8.3 Finding Unknown Angles

1. Answer: B
Explanation: 188-166=22

2. Answer: 90
Explanation: Adjacent angles share a vertex and a side. Add the measures of adjacent angles to find the total measure: 50+40=90.

3. Answer: 41.8
Explanation: The two angles are adjacent angles. Subtract 40.6 from 82.4.

4. Answer: D
Explanation: 72+44=116

5. Answer: C
Explanation: 134-88=46

6. Answer: 100°
Explanation: The angles are adjacent angles. Add the measures of adjacent angles to find the total measure: 40 + 60 = 100.

7. Answer: 34°
Explanation: The two angles are adjacent angles. To find the measure, subtract 34 from 68 degrees.

8. Answer: D
Explanation: The pizza is in the shape of a circle, which has 360 degrees. Each slice of pizza is 360/7, a degree. Dividing 360/7 by 2 gives the measure of half of the shared slice.

9. Answer: 43°
Explanation: The two angles are adjacent angles. Find the measure of the angle by subtracting: 82-39=43.

ANSWERS AND EXPLANATIONS

10. Answer: 19°
Explanation: The angles are adjacent angles. Add the measures of adjacent angles to find the total measure:
15+4=19

11. Answer: 34°
Explanation: The two angles are adjacent angles. Find the measure of the angle by subtracting:
52-18=34

12. Answer: D
Explanation: The two angles are adjacent angles. Find the measure of the angle by subtracting: 111-22=89

13. Answer: C
Explanation: The angles are adjacent angles. Add the measures of adjacent angles to find the total measure:
70+73=143

14. Answer: 12°
Explanation: The two angles are tadjacent angles. Find the measure of the angle by subtracting:
52-40=12

15. Answer: 118°
Explanation: The angles are adjacent angles. Add the measures of adjacent angles to find the total measure:
33+85=118

8.4 Chapter Review

1. Answer: D
Explanation: The angle is slightly less than a right angle (90 degrees) or 90/360 of the circle.

2. Answer: 360°
Explanation: An angle that turns through an entire circle has a measure of 360 degrees.

3. Answer: 180°
Explanation: An angle that turns through half a circle has a measure of 180 degrees.

4. Answer: B
Explanation: The angle has a measure of 56 degrees. An angle twice as large has as a measure of 56 2, or 112 degrees.

5. Answer: 94 degrees
Explanation: The angles are adjacent angles. Add the measures of adjacent angles to find the total measure:
22+72=94

6. Answer: 45°
Explanation: The angles are adjacent angles. Find the measure by subtracting:
85 − 40 = 45.

7. Answer: 60°
Explanation: The angles are adjacent angles. Find the measure by subtracting:
80 − 20 = 60.

ANSWERS AND EXPLANATIONS

8. Answer: 37°
Explanation: The corner of the farm is a 90-degree angle. The two fences create a 28-degree and 25-degree angle (28 + 25 = 53), Subtract 53 from 90.

9. Answer: $\frac{40}{360}$ or $\frac{1}{9}$

Explanation: A 40-degree angle turns through $\frac{40}{360}$ of the circle.

10. Answer: D
Explanation: The angle is approximately half of the right angle and turns through about $\frac{1}{8}$ of the circle. The fraction $\frac{40}{360}$ represents this value.

11. Answer: 60°
Explanation: The first angle has a measure of 20 degrees. Jack draws a total of 3 angles of that size. They have a combined measure of 20×3 or 60 degrees.

12. Answer: 1080°
Explanation: The wheel turns 360 degrees in 1 second. The wheel will turn 360 × 3 or 1080 degrees in 3 seconds.

13. Answer:

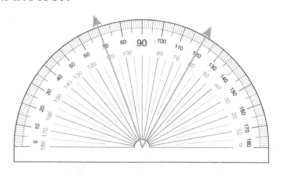

Explanation:
The first ray is at the 60-degree mark (or 120-degree mark). Using the markings on the inside of the protractor, add 60 and 50 to determine the angle of the second ray, the 110-degree marking. Or using the markings on the outside of the protractor, subtract 50 from 120 to determine the angle of the second ray, the 70-degree marking.

14. Answer: 45 and 140-degree, 40 and 145-degree
Explanation: Using the inside scale of the protractor, notice that the 2 rays are at the 45 and 140-degree markings. Using the outside scale of the protractor, notice that the 2 rays are at the 40 and 145-degree markings.

15. Answer: No
Explanation: The measure of Angle WXZ is 120 degrees. Write an equation: 120 = 52 + 2(4 × 3). Simplifying the right side gives 76. Thus, the measure is not 120 degrees.

ANSWERS AND EXPLANATIONS

16. Answer: 81°
Explanation: calculate $20 + (3 \times 2) + 110/2$. Following correct order of operations, the expression changes to $20 + 6 + 55$. Combine the terms with a result of 81.

17. Answer: Acute
Explanation: Angle, all three angles have measures that are less than 90 degrees.

18. Answer: Right
Explanation: The triangle has one angle that is exactly 90 degrees.

19. Answer: Right
Explanation: A right-angle has a measure of exactly 90°.

20. Answer: Acute
Explanation: An acute angle has a measure that is less than 90 degrees, but more than 0 degrees.

9. GEOMETRY

9.1 Angles and Sides of Quadrilaterals And Triangles

1. Answer: C
Explanation: A parallelogram has two pairs of opposite sides that are the same length. The markings show that the opposite sides have the same length.

2. Answer: B
Explanation: An isosceles triangle has two sides of the same length. It also has base angles with the same measure. The markings show that the two sides have the same length.

3. Answer: A
Explanation: Four angles in the quadrilateral are marked as right angles.

4. Answer: C
Explanation: All sides are a different length and all angles are a different measure. This describes a scalene triangle.

5. Answer: A
Explanation: A kite has two adjacent pairs of congruent sides.

6. Answer: C
Explanation: An obtuse triangle has only one angle that measures more than 90 degrees.

7. Answer: B
Explanation: The triangle with 3 equal sides that is also equiangular is an equilateral triangle.

8. Answer: D
Explanation: A trapezoid is a quadrilateral with one pair of parallel sides.

ANSWERS AND EXPLANATIONS

9. Answer: C
Explanation: A triangle with an interior angle of 90 degrees is a right triangle.

10. Answer: A
Explanation: The shape has four sides. It is a quadrilateral.

11. Answer: D
Explanation: An acute triangle is a figure where all three angles measure less than 90°.

12. Answer: B
Explanation: A triangle with one right angle and **two** acute angles is called a right triangle.

13. Answer: C
Explanation: A rhombus has 4 congruent sides and opposite, congruent angles.

14. Answer: A
Explanation: A square and a rhombus each have 4 congruent sides. A square has 4 congruent angles, and a rhombus has opposite congruent angles.

15. Answer: D
Explanation: The angle bisectors of a kite are perpendicular to each other.

9.2 Parallel and Perpendicular Lines of Quadrilaterals and Triangles

1. Answer: B
Explanation: A trapezoid has one set of parallel sides.

2. Answer: D
Explanation: By the definition of parallel lines and the definition of a triangle, a triangle cannot have any parallel lines.

3. Answer: C
Explanation: A right triangle has one right angle formed by perpendicular lines.

4. Answer: B
Explanation: A parallelogram must not have any perpendicular lines (right angles), but must have 2 pairs of parallel lines and opposite, congruent angles. So, it is false.

5. Answer: A
Explanation: A rectangle has two sets of parallel, congruent sides and 4 right angles.

6. Answer: C
Explanation: A quadrilateral is a 4-sided polygon. It does not have any parallel lines.

7. Answer: A
Explanation: Kite: Adjacent sides are equal. One pair of opposite angles is equal.

ANSWERS AND EXPLANATIONS

8. Answer: D
Explanation: A trapezoid is a quadrilateral with one pair of parallel sides.

9. Answer: D
Explanation: A parallelogram has two sets of parallel sides and opposite congruent angles.

10. Answer: B
Explanation:

A trapezoid has 1 pair of parallel lines. So, it is false

11. Answer: A
Explanation: The shape has 5 angles in the interior. All of the interior angles have measures that are greater than 90 degrees and less than 180 degrees.

12. Answer: B
Explanation:

There are 3 pairs of parallel lines in this polygon.

13. Answer: A
Explanation: A rectangle has 4 perpendicular lines which create the 4 right angles inside the rectangle. So, it is true.

14. Answer: C
Explanation: A rhombus with perpendicular sides would have right angles in all corners. This means the rhombus would have 4 congruent sides and 4 right angles making it a square.

15. Answer: A
Explanation: By the definition of a kite, the angle bisectors are perpendicular.

> **9.3 Lines of Symmetry**

1. Answer: A
Explanation: A line of symmetry divides a shape into two regions that are mirror images of each other across the line.

2. Answer: B
Explanation: A line of symmetry divides a shape into two regions that are mirror images of each other across the line. The line does not divide the candy into 2 mirror images.

3. Answer: C
Explanation: A scalene triangle has no lines of symmetry.

4. Answer: B
Explanation: A line of symmetry divides a shape into two regions that are mirror images of each other across the line. The line does not divide the apple into 2 mirror images.

ANSWERS AND EXPLANATIONS

5. Answer: A
Explanation: A line of symmetry divides a shape into two regions that are mirror images of each other across the line.

6. Answer: C
Explanation: The letter **M** has a vertical line of symmetry through the middle.

7. Answer: B
Explanation: A line of symmetry divides a shape into two regions that are mirror images of each other across the line. The dotted line does not divide the shape into identical, mirrored halves.

8. Answer: A
Explanation: A line of symmetry divides a shape into two regions that are mirror images of each other across the line.

9. Answer: A
Explanation: The letter **I** has horizontal and vertical lines of symmetry. So, **"I"** has 2 lines of symmetry. So, it is true.

10. Answer: C
Explanation: The letter O has a horizontal and vertical line of symmetry.

11. Answer: B
Explanation: The dotted line does not divide the shape into identical, mirrored halves.

12. Answer: A
Explanation: The dotted line divides the shape into identical, mirrored halves.

13. Answer: C
Explanation: The letter **X** has a horizontal line of symmetry through the middle.

14. Answer: A
Explanation: The letter **T** has a vertical line of symmetry through the middle. So, it is true.

15. Answer: A
Explanation: The dotted line divides the shape into identical, mirrored halves.

9.4 Chapter Review

1. Answer: D
Explanation: An obtuse triangle has exactly one angle that measures more than 90 degrees.

2. Answer: A
Explanation: By the definition of parallel lines and the definition of a triangle, an acute triangle cannot have any parallel lines.

3. Answer: A
Explanation: A line of symmetry divides a shape into two regions that are mirror images of each other across the line.

ANSWERS AND EXPLANATIONS

4. Answer: B
Explanation: The given shape has 2 sets of opposite congruent sides, so it is a parallelogram.

5. Answer: B
Explanation: The dotted line does not divide the shape into identical, mirrored halves.

6. Answer: C
Explanation: A rhombus has two sets of parallel, congruent sides and opposite, congruent angles.

7. Answer: B
Explanation: A trapezoid has no perpendicular lines. So, it is false.

8. Answer: C
Explanation: A kite has two adjacent pairs of congruent sides.

9. Answer: B
Explanation: An equilateral triangle has 3 lines of symmetry: from each point to the middle of the opposite side.

10. Answer: A
Explanation: A rectangle has opposite, congruent, parallel sides and 4 right angles. A parallelogram has 4 parallel sides and opposite congruent angles. Since a rectangle has 4 parallel sides and opposite, congruent angles, it is also a parallelogram.

11. Answer: A
Explanation: A shape with 3 sides and all the three angles measure less than 90° is an acute triangle. So, it is true.

12. Answer: C
Explanation: The letter D has a horizontal line of symmetry through the middle.

13. Answer: B
Explanation: A scalene triangle has no congruent sides and no congruent angles.

14. Answer: A
Explanation: A regular pentagon has 5 lines of symmetry.

15. Answer: C
Explanation: A quadrilateral is a polygon with 4 sides.

16. Answer: D
Explanation: The sum of the interior angles of a triangle is 180.

17. Answer: B
Explanation: The letter X has a vertical and a horizontal line of symmetry.

18. Answer: B
Explanation: A quadrilateral is a four-sided shape. A trapezoid is a quadrilateral with one pair of parallel sides.

ANSWERS AND EXPLANATIONS

19. Answer: B
Explanation: The letter N has no line of symmetry.

20. Answer: Circle
Explanation:

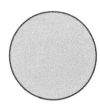

A circle is an example of a two-dimensional shape that has an infinite number of lines of symmetry.

Competitive Assessment 1
---

1. Answer: $100 = 5 \times 20$ or $100 = 20 \times 5$
Explanation: The equation shows that 5 groups of 20 equals 100 or 20 groups of 5 equals 100.

2. Answer: $A = 9 \times 6$
Explanation: The equation must use the variable A (to represent Adam), which shows that A is equal to 9 times 6.

3. Answer: B
Explanation: Express the answer by dividing 4,200 by 7.

4. Answer: Yes, he spent more than 5/7 of an hour practicing basketball.
Explanation: Compare the two fractions with the inequality $5/7 < 6/8$

5. Answer: 2,518 millimeters is more
Explanation: 1 m = 1,000 mm
2 m = 2,000 mm
So, 2,518 is more.

6. Answer: D
Explanation: Multiply the length times the height: $12 \times 8 = 96$.

7. Answer: C
Explanation: Calculate $11 + 33 = 44$. Then, $60 - 44 = 16$.

8. Answer: A
Explanation: Divide 91 by 13 which equals 7.

9. Answer: 45
Explanation: Calculate $15 \times 6 = 90$. Then, $90 - 45 = 45$.

10. Answer: B
Explanation: A round is 3,726 miles each way, which is 7,452 miles. There are 2 round trips, multiply 7,452 by 2 = 14,904

11. Answer: A
Explanation: If 6 out of 18 shoes are black and the rest are brown, subtract $18 - 6 = 12$ brown shoes. Thus, 12 out of 18 pairs are brown, which is equivalent to $\frac{6}{9}$ and $\frac{4}{6}$

$$\frac{12}{18} \div \frac{2}{2} = \frac{6}{9}$$

$$\frac{12}{18} \div \frac{3}{3} = \frac{4}{6}$$

ANSWERS AND EXPLANATIONS

12. Answer: $8\frac{11}{20}$
Explanation: Add the two thicknesses
$\frac{90}{60} + 1\frac{7}{20} = \frac{171}{20} = 8\frac{11}{20}$

13. Answer: 25×5=q
q=125
Explanation: The equation shows 25 groups of 5, which determines how many windows total there are in the school. There are 125 windows in total.

14. Answer: 6
Explanation: There are 96 pens, $\frac{96}{10} = 9$ with a remainder of 6. There are 6 pens left over.

15. Answer: No
Explanation: The number 6 is not a factor of 92. Dividing 92 by 6 leaves a remainder. Thus, 6 cannot be multiplied by another whole number to produce 92.

16. Answer: 27
Explanation: If half of the pencils are red, divide 104 by 2, giving 52 red pencils. Then, subtract 104 − 52 − 25 = 27. Of the pencils in the bag, 27 are black.

17. Answer: A
Explanation: The fraction of shirts that are black shirts is $\frac{7}{21}$ because 14 are white shirts and the rest are black shirts, 21−14=7. Then 7 out of 21 shirts are black shirts. The fraction $\frac{7}{21} = \frac{1}{3}$

18. Answer: No
Explanation: The number of oranges and mangoes in the basket is $\frac{80}{200}$ or $\frac{2}{5}$ which is less than half of the basket.

19. Answer: B
Explanation: A square has 4 sides.
17 squares have = 4×17 = 68

20. Answer: D
Explanation: Divide 522 by 16 which gives 32 with a remainder of 10.
Gil fills 32 boxes with 16 lemons each, and one box with 10 lemons.
So, he needs 33 boxes.

21. Answer: C
Explanation: Number of hamsters in pet shop=81
Number of rats in pet shop=9
Number of times more hamsters are there than rats =>81÷9=9

22. Answer: A
Explanation: Mariah bought 6 boxes of erasers.
Each box contains 24 erasers.
Number of erasers Mariah bought
=>6×24=y
y=144

23. Answer: D
Explanation: 64,898
Sixty four thousand eight hundred and ninety eight.

ANSWERS AND EXPLANATIONS

24. Answer: B
Explanation: JJaime buys 37 sunglasses. Each pair of sunglasses costs $14. Rounding the numbers 37 and 14 to the nearest 10.
40×10=$400

25. Answer: C
Explanation: A box contains 72 chocolates.
Number of boxes Hattie bought=6
Number of boxes Erin bought =11
Number of chocolates Hattie bought
=6×72=432
Number of chocolates Erin bought
=11×72=792
Number of chocolates both Hattie and Erin bought =432+792=1,224

26. Answer: A
Explanation: 27+4=31
31+4=35
35+4=39
39+4=43
43+4=47

27. Answer: 680
Explanation: The sequence starts with 320, so 320×2=640,
and 640−200=440
Then, 440×2=880, and 880−200=680

28. Answer: A
Explanation: The fraction $\frac{40}{480}$ is equivalent to $\frac{4}{48}$.
$\frac{4}{48} = \frac{1}{12}$ and $\frac{40}{480} = \frac{1}{12}$

29. Answer: B
Explanation: Add the numerators.
$\frac{9}{13} + \frac{7}{13} = \frac{16}{13}$ Do not change the denominators.

30. Answer: C
Explanation: 10:42 p.m−5:38 p.m
5 hours 4 minutes

31. Answer: 10 m 75 cm
Explanation: 4 m 20 cm+6 m 55 cm= 10 m 75 cm

32. Answer: A
Explanation: Add all the lengths of the 6 sides.

33. Answer: No
Explanation: The measure of the angle PQR is 170 degrees. Write an equation: 170 = 54 + 2 (4 X 7). Simplifying the right side gives 110. Thus, the measure is not 170 degrees.

34. Answer: No
Explanation:

ANSWERS AND EXPLANATIONS

35. Answer: 91 degrees
Explanation: Adjacent angles share a vertex and a side, but no interior points. Add the measures of adjacent angles to find the total measure:
57 + 34 = 91.

36. Answer: At the point O.
Explanation: Intersecting lines meet at the point O.

37. Answer: 25 degrees
Explanation: Adjacent angles share a vertex and a side. The measures of the two angled add
up to 76, so 76 − 51 = 25

38. Answer: B
Explanation: 11×8=88 cm².

39. Answer: 4×2 square inches
Explanation: There are 4 rows of squares, and each row contains 2 unshaded squares.
The area of the unshaded region is 4×2 square inches.

40. Answer: l=17 cm and P=64 cm
Explanation: Area, A = width×length
Substitute A=255 and w=15 in the equation.
255=15×l
l= $\frac{255}{15}$ =17 cm

Perimeter, P =2×(width + length)
Substitute l=17 and w=15 in the equation.
P=2×(15+17)=2×32=64 cm

41. Answer: z=129÷4
Explanation: The perimeter of a square is 129 inches. The length of its side is z inches.
P=4z
Therefore, 4z=129
(Or) z=129÷4.

42. Answer: A
Explanation: A line of symmetry divides a shape into two equal and symmetrical parts.

43. Answer: C
Explanation: Most people chose black.

44. Answer:

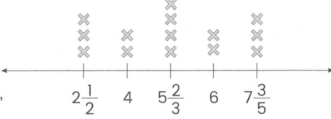

Length of Ribbons

Explanation: The data set has 14 values that match the numbers on the number line. Plot the
frequency of each value above the line in the appropriate place.

45. Answer: B
Explanation: The letter **H** has a horizontal and vertical line of symmetry. So, "H" has 2 lines of symmetry. So, it is False.

ANSWERS AND EXPLANATIONS

Competitive Assessment 2

1. Answer: 120=12×10 or 120=10×12
Explanation: The equation shows 120 as 10 times 12 or 12 times 10.

2. Answer: A
Explanation: Start with 7 times 24 and get 168. Then, 168 minus 94 is 74.

3. Answer: 22 bags
Explanation: The first step is to divide each original amount by the number of balls in each bag. Calculate $\frac{325}{14} = 23$ bags with 3 extra red balls, $\frac{265}{12} = 22$ bags with 1 extra blue ball, and $\frac{346}{10} = 34$ bags with 6 extra green balls. The number of bags is limited by the number of blue balls.

4. Answer: $\frac{40}{100} + \frac{80}{100}$ or $\frac{4}{10} + \frac{8}{10}$
Explanation: Converting the fraction to a common denominator before adding $\frac{40}{100}$ is equivalent $\frac{4}{10}$ and $\frac{8}{10}$ is equivalent to $\frac{80}{100}$.

5. Answer: 2 km
Explanation: 1 km=1,000 m
2 km=2,000 m
2,000>1,690

6. Answer: A
Explanation: Multiply the side length by itself: 6 × 6 = 36.

7. Answer: 2×5=10 or 5×2=10
Explanation: Two times 5 means that you should multiply 2 and 5, which equals 10.

8. Answer: 200,000
Explanation: Eight times means 25,000×8=200,000.
So people in amusement park=200,000

9. Answer: 6,000,000+500,000+ 20,000+1,000+400+20+7
Explanation: Expanding a number requires rewriting the number into a sum of every digit shown as its place value.

10. Answer: C
Explanation: Determine the speed by dividing 657 by 9.

11. Answer: $\frac{10}{12}$ cups of wheat flour and $\frac{2}{12}$ cups of butter
Explanation: Doubling the recipe means multiplying the ingredients by 2. When multiplying fractions, multiply the numerator by 2, but do not change the denominator.

12. Answer: A
Explanation: Add all the side lengths.
5+5+3+3=16

13. Answer: $41
Explanation: Multiply 2 times 41 equals 82 or divide 82 by 2 equals 41.

ANSWERS AND EXPLANATIONS

14. Answer: 11×10=110
Explanation: This story can be written as 11×10 = x because 11 groups of 10 is the same as 11 times 10.

15. Answer: B
Explanation: To find the maximum, multiply, 10 times 8 which equals 80.

16. Answer: 51
Explanation: If the rule is "add 4," then subtract 4 to find numbers in the front of the sequence. Thus, 67 minus 4 is 63, 63 minus 4 is 59, 59 minus 4 is 55, and 55 minus 4 is 51.
The sequence is 51, 55, 59, 63, 67,......

17. Answer: Seven million two hundred sixty-four thousand five hundred eighty-nine.
Explanation: Writing a number in written form is an expression of the number following base-ten numeral conventions.

18. Answer: A
Explanation: $\frac{8}{9}, \frac{5}{7}, \frac{1}{4}$

19. Answer: C
Explanation: One hour = $ 9
Per day 4 hours =4×$9=$36
One month =30×36=1,080

20. Answer: B
Explanation: One chocolate cake costs $13. Number of chocolate cakes bought for costs $208 =208÷13=16.

21. Answer: D
Explanation: Hope has 65 water bottles. Jada has 4 times as many water bottles as Hope =65×4=260.
Jada has 260 water bottles.

22. Answer: A
Explanation: Let r be the cost of one watermelon. Cost of twenty-two watermelons =$198 The cost of one watermelon =22×r=198
r=9

23. Answer: B
Explanation: Douglas thinking a number=18,426

24. Answer: C
Explanation: Nick sold 199 pies.
Each pie costs $8.
199×8=$1,592
$2,000 is the closest amount.

25. Answer: A
Explanation: Tanya's shop has 5,000 mushrooms.
Each box contains 25 mushrooms.
Each box costs $45.
5,000÷25=200
200×$45=$9,000

26. Answer: D
Explanation: The first two terms are 3 and 8
Third term =3+8=11
Fourth term =8+11=19
Fifth term =11+19=30

ANSWERS AND EXPLANATIONS

27. Answer: Multiply 2
Explanation: The pattern is multiplied by 2.

28. Answer: C
Explanation: The shaded portion represents the fraction $\frac{6}{12}$ which can be rewritten as the fraction 12. The fractions $\frac{1}{2}, \frac{12}{24}, \frac{18}{36}$ are equivalent to $\frac{6}{12}$

29. Answer: B
Explanation: Subtract the numerators. $\frac{14}{17} - \frac{10}{17} = \frac{4}{17}$ Do not change the denominators.

30. Answer: D
Explanation: 1 week=7 days
$7 \times 90 = \frac{630}{60} = 10.5$ hours

31. Answer: A
Explanation: 9 m=9×1,000=9,000 mm

32. Answer: $\frac{120}{360}, \frac{12}{36}, \frac{2}{6}$ or $\frac{1}{3}$

Explanation: A 120-degree angle turns through $\frac{120}{360}$ of the circle.

33. Answer: 9
Explanation: Count the number of data points to the left of 2. Nine people drink juice less than 2 cups of orange juice.

34. Answer:

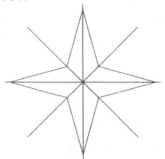

Explanation: A line of symmetry divides a figure into congruent parts that are mirror images of each other.

35. Answer: 115 degrees
Explanation: The measure of the angle can be determined using this expression $23+23\times4 = 115$

36. Answer: 163 square units.
Explanation: The area of a larger rectangle is 27×7=189 square units. The area of a small rectangle is 13×2=26 square units.
The area of the shaded shape is 189−26=163 square units.

37. Answer: B
Explanation: The shape is divided into three rectangles. Their areas are:
6×6=36 square centimeters
10×10=100 square centimeters
6×6=36 square centimeters
The area of the shape is 36+100+36=172 square centimeters.

ANSWERS AND EXPLANATIONS

38. Answer: D
Explanation: The area of a rectangle is given by A = width × length.
48=6×w
w= $\frac{48}{6}$ =8 yards.

39. Answer: A
Explanation: P=4a
We know that the perimeter is 60 inches.
So, 60=4a
a=60÷4
a=15 inches
Length of each side is doubled,
So, s=2×15=30 inches
New area, A=30×30=900 square inches.

40. Answer: C
Explanation: The height of the column that corresponds to Hamburger is 90.

41. Answer: A
Explanation: Area of the school=1800 m^2;
Length of the school=100 m
The area of the school hostel is the same as the school.
So, area of the hostel, A=1800 m^2
Length of the hostel, l=2×100=200 m
We know the area, A=l×w
Substitute A=1800 and l=200 in the equation.
Width of the hostel,
w= $\frac{A}{l}$ = $\frac{1800}{200}$ =9m

Perimeter, P=2(l+w)
Substitute l=200 and w=9 in the equation.
P=2(200+9)
P=2(209)=418 m.

42. Answer: B
Explanation: A line of symmetry divides a shape into two regions that are mirror images of each other across the line. The line does not divide the table into 2 mirror images.

43. Answer: C
Explanation: The letter O has infinite lines of symmetry.

44. Answer: 3
Explanation: There are 3 parallel lines in the hexagon.

45. Answer: B
Explanation: Option B is a rhombus.

Made in the USA
Las Vegas, NV
17 October 2023